PACEMAKER®

Practical
Mathematics
for Consumers

GLOBE FEARON

Pearson Learning Group

Pacemaker® Practical Mathematics for Consumers, Third Edition

We thank the following educators, who provided valuable comments and suggestions during the development of this book:

CONSULTANTS
Martha C. Beech, Center for Performance Technology, Florida State University, Tallahassee, Florida
Larry Timm, Special Education Teacher, Central Middle School, Midland, Michigan

REVIEWERS
Michael Brown, Department Chair, Mathematics, Hogan High School, Vallejo, California
Catherine Buckingham, Special Education Teacher, Friendly High School, Fort Washington, Maryland

PROJECT STAFF
Art and Design: Evelyn Bauer, Susan Brorein, Joan Jacobus, Jenifer Hixson, Jennifer Visco
Editorial: Jane Books, Danielle Camaleri, Phyllis Dunsay, Elizabeth Fernald, Dena Kennedy
Manufacturing: Mark Cirillo *Marketing:* Clare Harrison *Publishing Operations:* Travis Bailey, Debi Schlott

Photo Credits appear on page 371.

About the Cover: People need many skills to live successfully on their own. The images on the cover represent some of these skills. People need to earn money and stay on a budget. The circle graph helps you to organize all the expenses in your budget. You can see that rent is the greatest part of your total expenses. A checkbook helps you to manage your money. It is your own personal record of the money you deposit into an account and the checks you write to pay your expenses. A calculator helps you to calculate the balance in your checkbook accurately. To stay healthy, people need a balanced diet of nutritious food. Recreation is another important part of healthy lifestyle. What are some other things that are part of an independent lifestyle? How could these things be represented?

ISBN: 0-13-024146-6

Printed in the United States of America
10 11 12 13 V057 14 13 12 11

Globe Fearon
Pearson Learning Group

1-800-321-3106
www.pearsonlearning.com

Contents

A Note to the Student

In all areas of life, it's not what you know that counts—it's what you can *do* with what you know. Think about it. You probably already know how to add, subtract, multiply, and divide. But how well can you *apply* what you know to real-life problems and challenges?

That's what this book is all about. Practical Mathematics is math with a purpose. In these pages, you will find important information about buying, saving, and spending. You will learn how to read a paycheck stub and a checking account statement. You will learn how to make wise purchases when you shop for groceries, clothes, or furniture. By the time you finish this book, you will know about loans and interest, credit cards, and insurance. You will be able to make a budget, compare car prices, and plan a vacation you can afford. In fact, you will be able to "think math" when you need to make a plan or to get a job done. This ability will help you avoid many mistakes and problems. It will also help you make good decisions in all parts of your life—at school, at home, or on the job.

At the beginning of every unit, you will find a circle graph, a bar graph, or a line graph. You will be asked to read the graph and answer some questions. These graphs are similar to the kinds of graphs you find in newspapers and magazines.

At the beginning of every chapter, there is a list of **Learning Objectives**. They will help you focus on the important points covered in the chapter. **Words to Know** are also listed at the beginning of each chapter. They will give you a look ahead at the vocabulary you may find difficult. There is also a **Project**. This is an activity that is similar to the things you will need to do when you are living on your own. At the end of each chapter, a **Summary** will give you a quick review of what you just learned.

Throughout the book you will find notes in the margins of the pages. These friendly notes are there to make you stop and think. They comment on the material you are learning. Sometimes the notes give you tips. Sometimes they help explain a word.

Everyone who put this book together worked hard to make it useful, interesting, and enjoyable. The rest is up to you. We wish you well in your studies. Our success is in your accomplishments.

Making a budget helps you to manage expenses.

The circle graph shows expenses for one month. Use the graph to answer each question.

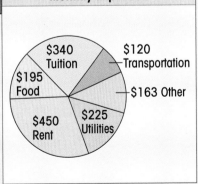

Monthly Expenses

$340 Tuition
$195 Food
$450 Rent
$225 Utilities
$120 Transportation
$163 Other

1. Which monthly expense is the greatest? The least?

2. How much does this person spend on food?

3. How much does this person spend on transportation?

4. Rent is the largest piece of the circle graph. What does this mean?

1

Covering Your Expenses

Living on your own means being able to pay your expenses. What are some expenses you will pay when you live on your own?

Learning Objectives

LIFE SKILLS

- Tell the difference between needs and wants.
- Prepare and analyze expense records.
- Identify variable and fixed expenses.
- Find the amount of income left after paying fixed expenses.
- Create a savings plan to pay for a big expense.

MATH SKILLS

- Add, subtract, multiply, and divide with money.
- Round money amounts to the nearest cent.

Words to Know

needs	things you must have, such as food
utilities	services such as electricity, gas, water, and heat
wants	things you would like, but can do without
expense	money spent on something you need or want
expense record	an organized chart used to list things you spent money on for a certain period of time, such as a week or a month
fixed expenses	expenses that are the same from week to week or month to month
variable expenses	expenses that change from week to week or month to month
income	money you earn for working, or receive from investments or other sources
savings	the difference between your income and your total expenses

Project: Moving Out on Your Own

You are moving out on your own. Look in a local newspaper or search the Internet to find the cost of renting a one-bedroom apartment in your area. You might find www.apartments.com to be a useful Web site. How much will your rent be? What else will you need? Make a list of things you will need to pay for and their costs. How much money will you need to cover all your expenses? Should you live alone or should you share an apartment?

Going out on your own means you have the freedom to make your own choices. Where will you live? What will you do?

Perhaps you are going to rent an apartment. If so, you will have certain **needs**. You will need **utilities**, such as gas, electricity, and water. You will also need food. You will have to pay for your needs.

You will also have certain **wants**. You may want to see a movie on a Saturday night. This is fine if you can afford it. However, seeing a movie is not something you need, like food and housing.

To understand needs and wants, Sandy made a list of what she spent for a week.

Did You Know?
What you *need* depends on your situation. If you ride your bicycle to work, repairing your bicycle is a *need*, not just a *want*.

MONEY SPENT WEEK OF APRIL 1ST

Pizza	$8.25	Bike repair	$10.49
Groceries	$65.00	Sweater	$40.65
Movie	$8.75	Nail polish	$4.25
Rent	$550.00	Toothpaste	$4.17

She then divided the list into two charts. On one chart, she included the items she needed. On the second chart, she included the items she wanted but did not really need.

Consumer Beware!
Advertisers will try to convince you that a *want* is a *need*. For example, a commercial might try to make you believe you need certain sneakers to be a good athlete; this is not true.

Need	Cost
Toothpaste	$4.17
Groceries	$65.00
Bike repair	$10.49
Rent	$550.00

Want	Cost
Movie	$8.75
Sweater	$40.65
Nail polish	$4.25
Pizza	$8.25

Practice and Apply

Use the charts on page 4 to solve each problem.

1. Which item that Sandy needed cost the least? Which cost the most?

2. Which items in Sandy's list cost more than the sweater?

3. List the items from the wants chart in order from least expensive to most expensive.

4. Why do you think pizza is listed as something that Sandy wanted rather than needed?

5. Name two items from Sandy's wants chart that she could have gone without to save $12.50 for next week's groceries.

Here's a Tip!
The difference between wants and needs isn't always clear. A warm, well-made sweater that costs $20 is a need if your others are old and worn. A sweater with a fancier design that costs $60 is a want.

Abraham made this list of what he spent last week. Use Abraham's list to solve each problem.

MONEY SPENT WEEK OF JUNE 7ᵗʰ

Bus pass	$42.00	Movie	$8.50
Groceries	$57.25	Medicine	$28.59
Shirt	$15.00	Software	$40.69
Popcorn	$2.79	Concert	$43.00

6. Make two charts, one for needs and one for wants. Then, fill in each chart, using the items from Abraham's list.

7. Which item that Abraham wanted cost the most? Which cost the least?

8. **CRITICAL THINKING** Is computer software a need or a want? Explain your thinking.

1·2 Focus: Adding and Subtracting With Money

It is important to keep track of your money. You will need to add or subtract to manage your money. An **expense** is money you spend on a need or a want. If you want to find the total of your expenses, you add.

▶ **EXAMPLE 1**

Aaron spent $6.50 on a round-trip bus ticket to go to work and home, $5.25 for lunch, and $0.75 on a snack. What were Aaron's expenses for the day?

STEP 1 First, write the numbers in vertical form. Line up the decimal points.

$$\begin{array}{r} \$6.50 \\ 5.25 \\ + \ 0.75 \end{array}$$

STEP 2 Add the digits in columns from right to left. Regroup as needed. Place the decimal point in the sum.

$$\begin{array}{r} \overset{1\ 1}{} \\ \$6.50 \\ 5.25 \\ + \ 0.75 \\ \hline \$12.50 \end{array}$$

Aaron's expenses for the day were $12.50.

If you want to find how much more one expense is than another, you subtract.

▶ **EXAMPLE 2**

The next day, Aaron spent $35.00 on items he needed. He also spent $18.75 on items he wanted. How much more did he spend on his needs than on his wants?

STEP 1 First, write the numbers in vertical form. Line up the decimal points.

$$\begin{array}{r} \$35.00 \\ - \ 18.75 \end{array}$$

STEP 2 Subtract the digits in columns from right to left. Regroup as needed. Place the decimal point in the difference.

$$\begin{array}{r} \overset{2\,14\,9\,10}{\$35.00} \\ - \ 18.75 \\ \hline \$16.25 \end{array}$$

Aaron spent $16.25 more on his needs.

Skills Practice

Add.

1. $721.38
+ 378.43

2. $72.95
+ 8.26

3. $21.62
    9.27
+ 0.14

4. $89.50
    7.35
+ 25.79

5. $101.99 + $450.00 + $0.07

6. $428.50 + $12.35 + $0.79

7. $14.38 + $165.00 + $5.98

8. $10.00 + $17.89 + $9.59

Subtract.

9. $287.56
− 38.27

10. $45.08
− 5.98

11. $50.00
− 35.98

12. $72.05
− 29.69

13. $101.05 − $22.03

14. $300.00 − $79.85

15. $5,000 − $199

16. $708.20 − $38.50

Everyday Problem Solving

Juan spent $11.20 on food. His receipt is on the right. Some of the prices are missing. Use the receipt to solve each problem.

1. What was the total cost of the bread, milk, and ham?

2. If Juan spent $2 on apples, how much did the cheese cost? (Hint: Find the sum of the items you know. Subtract it from the total.)

3. If Juan spent $3.50 on cheese, how much did the apples cost?

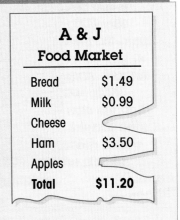

A & J
Food Market

Bread	$1.49
Milk	$0.99
Cheese	
Ham	$3.50
Apples	
Total	**$11.20**

What Is an Expense Record?

Keeping a record of expenses helps you predict how much money you need each month. A list of expenses is called an **expense record**. Jake's expenses for May are listed below.

Week 1		Week 2		Week 3		Week 4	
Expense	**Cost**	**Expense**	**Cost**	**Expense**	**Cost**	**Expense**	**Cost**
Groceries	$46.42	Dinner	$13.75	Lunch	$3.50	Electricity	$42.95
Dry cleaning	$11.25	Jeans	$32.95	Socks	$3.50	Gasoline	$17.50
Lunch	$4.25	Gift	$25.00	Stamps	$5.00	Gift	$14.00
Gasoline	$10.00	Telephone	$15.82	Groceries	$15.36	Concert	$15.00
Parking	$3.00	Groceries	$16.37	Haircut	$15.00	Rent	$550.00
School book	$41.29	Car insurance	$77.00	Movie	$5.50	Groceries	$29.00

▶ **EXAMPLE**

Did Jake spend more money in week 1 or in week 2? How much more?

STEP 1 Add the expenses for each week.

Week 1 $46.42 + $11.25 + $4.25 +
$10.00 + $3.00 + $41.29 = $116.21
Week 2 $13.75 + $32.95 + $25.00 +
$15.82 + $16.37 + $77.00 = $180.89

Here's a Tip!
The symbol < means *is less than.*

STEP 2 Compare the expenses in week 1 to those in week 2.

$116.21 < $180.89

Jake spent more money in week 2 than in week 1.

STEP 3 Subtract to find how much more money Jake spent in the second week.

$180.89
− 116.21
$64.68

Jake spent $64.68 more in week 2.

Jake organizes weekly expense records into one monthly expense record. He places groceries, lunch, and dinner into a food category. He determines how much he spent for food each week.

Category	Week 1	Week 2	Week 3	Week 4	Total
Food	$50.67	$30.12	$18.86	$29.00	
Clothing					
Transportation					
Personal care					
Recreation					
Utilities					
Car insurance					
Rent					
Other					
Total					

Practice and Apply

Solve each problem. Show your work.

1. Copy the monthly expense record above. Use Jake's weekly expense records on page 8 to complete the record. Use the row labeled *Other* for expenses that do not fit in a category. Use a calculator if you like.

2. Look at the total column. For which category did Jake spend the most money? The least money?

3. How much less money did Jake spend on recreation than on food?

4. In which week were Jake's expenses the greatest?

5. How much were Jake's total expenses for the month?

6. **IN YOUR WORLD** Keep a record of your expenses for four weeks. Make a monthly expense record by category.

What Are Fixed and Variable Expenses?

Fixed expenses stay the same each week or month. Rent is a fixed expense. Fixed expenses usually do not change.

Other expenses can change. These expenses are called **variable expenses.** For example, you probably spend a different amount for groceries each week. Food is a variable expense. You can control variable expenses.

The charts below list Amy's expenses for the month.

Fixed Expense	Cost
Rent	$650.00
Car insurance	$110.50
Car payment	$150.00
Club dues	$8.50
Newspaper subscription	$6.25
Internet connection	$23.30
School loan	$64.00

Variable Expense	Cost
Birthday gift	$13.00
Burger Barn dinner	$9.26
Clothing	$48.73
Groceries	$74.96
Telephone	$14.44
Utilities	$32.00
Haircut	$15.00
Movie	$10.00
Snacks	$22.50

Practice and Apply

Solve each problem. Show your work.

1. How much more did Amy spend on fixed expenses than on variable expenses?

2. Which variable expenses may not occur every month?

3. Amy has $1,350 to spend for the month. Add all her expenses. Will she have any money left over? How much?

4. **IN YOUR WORLD** Name two fixed expenses and two variable expenses that your family has each month.

What Is Income?

Jill pays for her expenses by working at a pet clinic. The money she earns plus any other money coming in is her **income**. The difference between her income and her total expenses is the amount she has left for savings.

▶ **EXAMPLE**

Jill earns $1,350 each month. Her total expenses for each month are $1,012.55. How much money does Jill have left for savings?

Here's a Tip!
You can open a savings account at a bank with the money you save.

$$\begin{array}{rl} \$1,350.00 & \text{income} \\ - \quad 1,012.55 & \text{expenses} \\ \hline \$337.45 & \text{money left} \end{array}$$

Jill has $337.45 left for savings.

Sometimes you do not spend all of your income. **Savings** is the difference between your income and the total of your fixed and variable expenses.

Practice and Apply

Complete the chart.

	Occupation	Monthly Income	Total Expenses	Money Left for Savings
1.	Security Guard	$1,480	$989	?
2.	Stock Clerk	$832	$668	?
3.	Fast-Food Worker	$722	$617	?
4.	Factory Worker	$1,180	$978	?
5.	Sales Associate	$787	$656	?

6. **WRITE ABOUT IT** Maya earns $1,327 a month. Last month her fixed expenses were $740 and her variable expenses were $389. How much did she save? Explain.

Focus: Multiplying and Dividing With Money

Sometimes you may need to multiply or divide with money.

► EXAMPLE 1

Logan earns $468 a week. How much does he earn in a year? (Hint: There are 52 weeks in a year.)

Here's a Tip!
Multiply by 52 to find how much Logan earns in a year.

STEP 1   Multiply. First, multiply 468 by 2 ones. Regroup as needed.

$$\begin{array}{r} {}^{1\ 1} \\ \$468 \\ \times\ \ 52 \\ \hline 936 \end{array}$$

STEP 2   Multiply 468 by 5 tens. Regroup as needed.

$$\begin{array}{r} {}^{3\ 4} \\ {}^{\cancel{1}} \\ \$468 \\ \times\ \ 52 \\ \hline 936 \\ 23400 \end{array}$$

STEP 3   Add. Regroup as needed.

$$\begin{array}{r} {}^{3\ 4} \\ {}^{\cancel{1}} \\ \$468 \\ \times\ \ 52 \\ \hline 936 \\ +\ 23\ 400 \\ \hline \$24{,}336 \end{array}$$

Logan earns $24,336 a year.

► EXAMPLE 2

How much does Logan earn each month? (Hint: There are 12 months in a year.)

Here's a Tip!
Divide by 12 to find how much Logan earns each month.

STEP 1   Divide. First, divide the thousands. Write 2 in the quotient for 2 thousands. Multiply 2 × 12. Subtract. Bring down the next digits.

$$\begin{array}{r} \$\ \ 2\ \ \ \ \ \\ 12\overline{)\$24{,}336} \\ -\ 24\downarrow\downarrow \\ \hline 0\ 33 \end{array}$$

STEP 2   Divide the hundreds. There are not enough hundreds. Write 0 in the quotient for 0 hundreds. Divide 33 tens. Multiply 2 × 12. Subtract. Bring down the next digit.

$$\begin{array}{r} \$\ \ 2{,}02\ \ \\ 12\overline{)\$24{,}336} \\ -\ 24 \\ \hline 0\ 33 \\ -\ 24\downarrow \\ \hline 96 \end{array}$$

STEP 3   Divide 96 ones. Write 8 in the quotient for 8 ones. Multiply 8 × 12. Subtract. There is no remainder.

$$\begin{array}{r} \$\ \ 2{,}028 \\ 12\overline{)\$24{,}336} \\ -\ 24 \\ \hline 0\ 33 \\ -\ 24 \\ \hline 96 \\ -96 \\ \hline 0 \end{array}$$

Logan earns $2,028 each month.

Skills Practice

Multiply.

1. $412
    × 35

2. $3.62
    × 21

3. $37.80
    × 52

4. $4,782
    × 48

5. $905
    × 75

6. $159.68
    × 12

7. $234 × 32

8. $6.15 × 40

9. $12.00 × 52

Divide.

10. 15)$3,975

11. 13)$4,680

12. 23)$47,150

13. 35)$301

14. 35)$30,100

15. 11)$2,222

16. $58,240 ÷ 52

17. $6,516 ÷ 12

18. $10,890 ÷ 36

Everyday Problem Solving

Use the newspaper advertisement Brian found to solve each problem.

1. How much will the rent for the apartment be for one year?

2. Last year the utilities cost $1,224 for the year. How much were the utilities per month?

3. Brian's monthly income is $1,850. If he pays his rent and spends $39 on the telephone, $276 on groceries, and $102 on utilities, how much money will he have left?

APARTMENT FOR RENT

$475 per month, utilities not included.
(848) 555-8746

1·7 How Do You Save for a Big Expense?

What do you do when you want to buy something that you cannot afford? You don't have enough money left over after paying your monthly expenses to make your purchase. You need to create a savings plan.

EXAMPLE 1

You plan to save $22 each month to buy a DVD player. The DVD player costs $220. Will you be able to buy it in 6 months? If not, how many months do you need to save for the DVD player?

STEP 1 Multiply to find out how much you can save in 6 months.

$$\begin{array}{r} \$22.00 \\ \times 6 \\ \hline \$132.00 \end{array} \quad \begin{array}{l} \text{savings each month} \\ \text{number of months} \\ \text{total amount saved} \end{array}$$

STEP 2 Compare the amount saved to the amount needed.

$132 < $220

You won't have enough money in 6 months to buy the DVD player.

STEP 3 Divide to find how many months you need to save for the DVD player.

$$\begin{array}{r} 10 \\ \$22\overline{)\$220.00} \\ -22 \\ \hline 0 \end{array}$$

You need to save $22 for 10 months.

Sometimes you want to buy something by a certain time. You want to know how much to save each month.

EXAMPLE 2

The DVD player costs $220. You want to buy the DVD player in 6 months. How much do you need to save each month?

STEP 1	Divide the money needed by the number of months.	$36.66... 6)$220.00 − 18 40 − 36 40 − 36 40 − 36 4

Here's a Tip!
You can round to the nearest dollar. Look at the first digit after the decimal point. Round up if it is 5 or higher. Round down if it is 4 or lower.

STEP 2 Round up to the nearest dollar. $36.66... ⟶ $37

You would have to save about $37 each month.

Practice and Apply

Solve each problem. Show your work.

1. Suppose you want a guitar. It costs $756. About how much would you have to save every month for 6 months to be able to buy the guitar? For 7 months?

2. The new set of tires you will need for your car costs $320. You have $80 saved. How much will you need to save each month to buy the tires in 3 months?

 3. You want a $300 couch as soon as possible. How much would you have to save every month for 5 months to afford the couch? For 4 months? For 3 months? Use a calculator if you like.

Maintaining Skills

Compute.

1. $48.00 − $37.98   **2.** $327 ÷ 3   **3.** $325 + $869   **4.** $510 × 12

5. $35 × 7   **6.** $60 ÷ 12   **7.** $416 − $199   **8.** $732 + $599

Solve each problem. Show your work.

1. Jerome has a job that requires a clean shirt every day. It costs $1.25 to have a shirt washed and ironed. What is the cost for clean shirts for 5 days?

2. If Martina takes the bus to work, she pays $15.00 a week. If she car pools, she pays $7.75 a week. How much does she save if she car pools for the week?

3. If Duane packs his lunch, it costs about $10.50 for five days. If he buys lunch at work, it costs about $3.50 a day. About how much does he save each week by packing a lunch? (Hint: First find how much it costs to buy lunch for 5 days.)

4. Shirley decides to buy a weekly train pass for $25 instead of paying $3.25 each way. She takes the train to work and back home again for 5 days. How much money does she save with the weekly pass?

5. **OPEN ENDED** Stephanie takes a bus and a train to work every day. It costs her a total of $5 one way. What might be the cost of the bus? The train?

Calculator

Be careful when you use a calculator to divide. Always enter the dividend first. The dividend is the number being divided.

$25\overline{)150}$ is the same as $150 \div 25$. Enter 150 first.

Use a calculator to find the value of each expression.

1. $129 \div 3$ 2. $14\overline{)\$355.88}$ 3. $305.24 \div 4$ 4. $9\overline{)\$163.08}$

5. $42\overline{)\$756}$ 6. $31.50 \div 0.25$ 7. $50\overline{)\$1,650}$ 8. $1.48 \div 2$

DECISION MAKING:
How Could You Cut Expenses?

Jeff is a 28-year-old independent taxicab driver. He is living alone in a two-bedroom apartment. Jeff works four days a week. His income after taxes is about $565 each week. Jeff must make his income cover all his monthly expenses.

Next month, Jeff has additional expenses that do not fit into his budget. He must rent a tuxedo and buy a gift for his brother's wedding. These things will cost him $200.

Jeff organized his monthly expenses into one expense record.

Use the expense record to solve each problem.

1. Which items do you think Jeff could do without for one month? Suppose he did not spend money on these items for one month. Would he have the $200 he needs for these additional expenses?

2. What long-term changes could Jeff make in his life that would allow him to save more money each month?

You Decide

Jeff has increased the number of hours he works each week. He now makes about $600 each week. What should Jeff do with the extra income? Why?

Expense	Cost
Rent	$950.00
Utilities	$170.67
Phone	$39.33
Cable	$35.00
Groceries/ personal care	$321.12
Gasoline	$297.64
Clothes	$71.24
Car insurance	$150.00
Miscellaneous: candy, magazine, etc.	$34.37
Haircut	$20.00
Recreation	$98.75
CDs	$26.73
Dining out	$45.15
Total	$2,260.00

Summary

It is important to tell the difference between things that you want, such as concert tickets and CDs, and things that you need, such as food and soap.

You can keep track of where your money goes by keeping an expense record.

A fixed expense does not usually change from month to month.

A variable expense is one that you can change from month to month.

The money you receive is your income. Your expenses are what you pay.

After you pay for expenses, you can save what is left.

If you want something costly, you can use a savings plan to decide how much to save each month.

expense record

expense

fixed expenses

income

needs

savings

utilities

variable expenses

wants

Vocabulary Review

Complete the sentences with words from the box.

1. Money you earn for working, or receive from investments or other sources, is ____.

2. Things you must have, such as food, are ____.

3. When you spend money on something that you need or want, you are paying an ____.

4. An ____ is an organized chart used to list things you spend money on for a certain period of time, such as a week or a month.

5. ____ change from week to week or month to month.

6. Things you would like but can do without are ____.

7. ____ are the same from week to week or month to month.

8. Services such as electricity, gas, and water are ____.

9. The difference between your income and your total expenses is ____.

Chapter Quiz

Use the expense record at the right to solve each problem.

Tracy's Expense Record	
Expense	Cost
Rent	$700.00
Food	$86.19
Clothing	$45.87
Gasoline	$34.00
Personal care	$24.00
Recreation	$12.50
Gifts	$5.75
Utilities	$37.50
Car insurance	$90.00
Car loan	$100.00
Internet	$25.00

1. What was Tracy's greatest expense for the month?

2. Tracy has seven variable expenses. List these expenses and find their total.

3. How much more did she spend on food than on clothing?

4. List Tracy's fixed expenses. Find the total.

5. Which cost more — her fixed expenses or her variable expenses?

6. What is the total of Tracy's expenses?

7. Tracy's income is $1,194.83 a month. Is that enough to cover her total expenses? If so, how much does Tracy have left for savings?

8. If your income were $998.00, would you be able to afford the same expenses as Tracy? If not, how much more would you need?

Maintaining Skills

Compute.

1. $405.75
  + 395.25

2. $1,200
  − 869

3. $215.50
  × 35

4. $1,325
  + 439

5. $300.50
  × 8

6. $5,209
  − 345

7. 5)$1,650

8. $19.99
  + 0.75

You will want to have enough money to buy the things you need. Make a budget that shows your income and expenses. What are some things you need to buy?

Learning Objectives

LIFE SKILLS

- Prepare a budget.
- Find ways to make your income meet your needs.
- Determine steps you can take to balance a budget.
- Determine ways to cope with financial emergencies.
- Decide how to budget and spend extra money.

MATH SKILLS

- Estimate costs to the nearest dollar.
- Find an average.
- Add, subtract, multiply, and divide with money.

Words to Know

budget	a plan for spending and saving money
transfer	to move money from one budget line to another
balanced budget	a plan where income is equal to the sum of expenses and savings
estimate	to round to get an answer that is not exact; the answer you get when you use rounded numbers
interest	extra money you need to pay the lender for loaning you money
predict	to make a guess about the future
debt	money that is owed
average	the sum of a set of numbers divided by how many numbers are in the set

Project: Researching a Budget

Find a copy of a budget for your school district, town, or state. These budgets can be found at your town's library, at the town hall, or on the Internet. Two useful Web sites are www.nasbo.org, the National Association of State Budget Officers, and www.fedstats.gov, which has statistics for more than 100 federal agencies.

Review the budget. Then answer the following questions.

- What is the total of all expenses?
- Where is the greatest amount of money spent?
- Where is the least amount of money spent?

What's in a Budget?

Wordwise
The word *budget* comes from the old English word *bouget*, meaning *wallet*.

A **budget** is a plan for what to do with your income. A budget consists of three parts. The first part is income. The second part is expenses. Expenses include fixed expenses and variable expenses. The third part is savings. A budget can help you see what you can and cannot afford.

Suppose you need an apartment. Should you live alone or with a roommate? Look at your expenses to help you decide.

► **EXAMPLE 1**

Gloria's income is $1,250 a month. She lives with a roommate. She splits these expenses: rent $880, phone $40, utilities $128, and cable TV $32. What is Gloria's share of these expenses?

STEP 1   Add the expenses.

$880	rent
40	phone
128	utilities
+ 32	cable TV
$1,080	

STEP 2   Divide the sum by 2.

$$\begin{array}{r} 540 \\ 2\overline{)\$1,080} \end{array}$$

Gloria's share of these expenses is $540.

If Gloria lived alone, she would have to pay the total $1,080 for these expenses. She would have very little money left over. By having a roommate she has more money left for other expenses and savings.

► **EXAMPLE 2**

How much does Gloria have left for other expenses and savings?

Subtract Gloria's share of the expenses from Gloria's income.

$1,250
−   540
$710

Gloria has $710 left for other expenses and savings.

Practice and Apply

Sam shares an apartment with two friends. His monthly income is $1,500. These are his shared monthly expenses.

Sam's Shared Expenses			
Fixed Expense	Cost	Variable Expense	Cost
Rent	$800.00	Utilities	$50.00
Cable TV	$45.00	Phone	$35.00

1. Sam's share of the utilities is about $50 a month. Each of his roommates pay the same amount. How much is the total utility bill?

2. What is the total of Sam's monthly fixed expenses for the apartment?

3. After Sam pays his shared expenses, how much money does he have left for savings and other expenses?

4. If Sam saves $300 a month, how much does he have left to spend on other expenses?

5. Sam wants to buy a used car. He wants to spend about $2,100 on the car. If he saves $300 a month, how many months will he need to save to buy the car?

6. Suppose the total rent for the apartment is increased by $150. How much would Sam's share be then?

7. **WRITE ABOUT IT** Suppose another person moved in with Sam and his roommates. Would Sam's fixed apartment expenses stay the same? Explain why or why not.

What if you have a new expense? How would you pay for it? You need to change your budget. You do this by subtracting money from savings and variable expenses. Then you **transfer** the money you subtract to cover the new expense. Remember that you always need to keep a **balanced budget**. You can't spend and save more money than your income.

▶ **EXAMPLE 1**

This is Kelly's monthly budget.

Kelly's Monthly Budget

Monthly Income			$1,800
Expenses			
Rent	$760	Clothing	$90
Utilities	$146	Laundry	$30
Phone	$25	Recreation	$45
Groceries	$211	Bus fare	$43
Savings			$450

Here's a Tip!
When deciding which budget items to give up or reduce, always choose to give up or reduce wants. Needs are essential to living well.

Kelly needs a washer and a dryer. She cannot pay for these at once. So, she decides to pay $60 a month for a year. Her utilities increase $15 a month. Her laundry expenses decrease $30 a month. Will her expenses increase or decrease? How much?

		Expenses Added	Expenses Subtracted
STEP 1	Add the new expenses to the budget. Then compare that amount to the expenses subtracted from the budget.	$60 + 15 $75	$30

$$\$75 > \$30$$

Expenses added are greater than expenses subtracted. So, there will be an increase in total expenses.

STEP 2 Subtract to find the amount that expenses increased.

$$\$75 - \$30 = \$45$$

There will be a $45 increase in expenses.

► **EXAMPLE 2**

Kelly will transfer money from savings to cover the new expenses. Show the transfers in Kelly's budget.

Add the $60 expense for the washer and dryer.
Add $15 to utilities.
Subtract $30 from laundry.
Subtract $45 from savings.

Kelly's Monthly Budget

Monthly Income			$1,800

Expenses

Rent	$760	Clothing	$90
Utilities	~~$161~~ $146	~~Laundry~~ ~~$30~~ $0	
Phone	$25	Recreation	$45
Groceries	$211	Bus fare	$43
Washer/Dryer	$60		

Savings			~~$450~~ $405

The transfers to Kelly's budget are shown in blue.

Practice and Apply

Use Mike's monthly budget to solve each problem.

1. Mike must add $20 to clothing, $7 to bus fare, and $5 to Internet. Is there enough money in recreation to transfer for these expenses? If yes, how much will be left?

Mike's Monthly Budget

Monthly Income			$334

Expenses

Dining out	$98	Bus fare	$56
Internet	$20	Recreation	$100
Clothing	$35		

Savings			$25

2. Make the transfers suggested in exercise 1. Show what Mike's new budget will look like.

3. **CRITICAL THINKING** Mike wants to save $50 more each month. Use Mike's original budget above. Transfer money to cover the additional savings.

Focus: Estimating With Money

Have you ever gone to the supermarket and worried if you had enough money to pay for your items? A quick way to find out is to estimate. You **estimate** when you don't need an exact answer. An **estimate** is found by rounding exact numbers.

EXAMPLE 1

Joe is at the supermarket. He has $18.42 in his wallet. He put the following items in his basket: bread for $1.39, milk for $1.89, carrots for $0.79, orange juice for $2.99, and cereal for $3.82. Estimate to see if Joe has enough money to pay for the groceries.

Here's a Tip!
It is a good idea to round up when estimating purchases. Then you can always be sure you have enough money.

STEP 1	Estimate. Round up to the next dollar. Then add.	$1.39	→	$2.00 bread
		1.89	→	2.00 milk
		0.79	→	1.00 carrots
		2.99	→	3.00 orange juice
		3.82	→	+ 4.00 cereal
				$12.00

Joe needs about $12.00 to pay for the groceries.

STEP 2 Compare the estimate with the amount Joe has to spend. $12.00 < $18.42

Joe has enough money to pay for the groceries.

Sometimes, it is best to round to the nearest dollar.

EXAMPLE 2

Wordwise
The word *about* is a clue to estimate.

The groceries cost exactly $10.88. Joe has $18.42 in his wallet. About how much money will Joe have left after buying groceries? (Hint: Find the difference.)

Estimate. Round to the nearest dollar. Then subtract.	$18.42	→	$18.00
	10.88	→	− 11.00
			$7.00

Joe will have about $7.00 left.

Estimate each sum. Round up to the nearest dollar. Then add.

1. $4.92
    + 3.44

2. $12.25
    + 19.31

3. $21.62
    + 20.49

4. $41.35
    + 6.23

Estimate each sum or difference. Round to the nearest dollar.
Then add or subtract.

5. $3.42
    + 1.98

6. $24.87
    + 14.58

7. $43.61
    − 15.67

8. $38.92
    − 7.15

9. $26.75 + $9.12 + $8.45

10. $395.12 + $46.98 + $3.02

11. $95.74 − $8.79

12. $110.20 − $12.08

Everyday Problem Solving

To solve each problem, round to the nearest
$10. Then add or subtract. Show your work.

1. Estimate how much Carmen spent on fixed
expenses. Then estimate how much Carmen
spent on variable expenses.

2. About how much more did Carmen spend
on fixed expenses than on variable
expenses?

3. Estimate Carmen's total expenses for her car.

4. Carmen wants to buy three books that cost
$9.75, $14.25, and $5.68. Will she be able to
buy all three if she has $27?

Carmen's Expenses for July	
Fixed Expenses	
Rent	$700
Cable TV	$45
Health Club	$25
Loan	$100
Car Loan	$129
Car Insurance	$100
Variable Expenses	
Utilities	$179
Gasoline	$42
Phone	$39
Sweater	$27
Manicure	$17
Groceries	$162

You have been saving $100 a month from your job at the factory to go to computer school. Now the computer school is offering just the right course for you.

You have saved $1,500. But the cost of the course is $2,500. If you borrow $1,000 from the bank, you will be charged **interest**. You decide to take out a loan. Your payments will be $120 a month.

Consumer Beware!
If you are thinking about attending a community college, compare costs. It usually costs less to attend a community college in your own county.

A new friend at work is also going to take the course. Luckily for you, your friend has a car. He will drive you to work and to school, and then bring you home after school. He would like you to pay $10 a week for gas money. That is better than the $15 a week you paid to ride the bus.

▶ **EXAMPLE**

Will going to computer school change the total expenses in your budget?

Budget Changes for Computer School	
Expenses Added to Budget	**Expenses Subtracted from Budget**
• $120 per month for loan • $40 per month for gas	• $100 per month savings for school • $60 per month for bus fare

STEP 1 Find the sum of the expenses added to your budget. $120 + $40 = $160

STEP 2 Find the sum of the expenses subtracted from your budget $100 + $60 = $160

STEP 3 Compare the numbers. $160 = $160

Going to computer school will not change the total of the expenses in your budget.

This is why a budget is important. Now it is clear that you can afford to take the computer class.

Practice and Apply

Lisa wants to buy a car. She listed the expenses she will add and subtract from her budget.

Budget Changes to Buy a Car	
Expenses Added to Budget	**Expenses Subtracted from Budget**
• $239 a month for a car loan • $40 per month for gas • $100 a month for car insurance	• $100 a month for train fare, bus fare, and taxicab rides • $150 a month savings for a car

1. What is the total of the expenses Lisa will add to her monthly budget if she buys the car?

2. What is the total of the expenses Lisa will subtract from her budget if she buys the car?

3. Which is greater, the expenses she will add or the expenses she will subtract? How much greater?

4. Car insurance costs $100 a month. How much does car insurance cost in a year?

 5. Lisa will pay $239 a month for the car loan. How much will she pay in 12 months? In 2 years? Use a calculator if you like.

6. **WRITE ABOUT IT** Would it be wise for Lisa to buy an old car? Why would the age of the car make a difference?

What About Changes You Can't Predict?

Something has happened that you couldn't **predict.** The apartment building you live in is being sold. You will have to move. Your roommate has decided to work in another city. You need to find a place of your own.

You find a studio apartment you like. Now you have to pay the total rent, gas, electricity, cable TV, and phone. The utility bills aren't as high as they were for two people, but they are more than you have been paying.

You get a $34 a month raise at your regular job, but this will not be enough. So, you get a second job. You work eight evening shifts each month at Don's Ice Cream Shop.

Here's how you balance your new budget.

Wordwise

A *studio apartment* is a small apartment. It usually has one main living space, a small kitchen, and a bathroom.

Budget Increases	
Increased Monthly Income	**Increased Monthly Expenses**
• $34 a raise at work • $168 for working at Don's Ice Cream Shop	• $133 rent increase • $32 utilities increase • $18 phone increase • $15 cable TV increase

Here's a Tip!

When estimating your total income, it's good to round down so you don't overestimate the amount of money you will have.
So, $168 → $160.

You can see that there are more increases in expenses than increases in income. However, your increased income may still total more than your increased expenses.

You can estimate if you want a quick comparison. To know for sure, you'll have to do the math.

Practice and Apply

Use the chart on page 30 to solve each problem. Show your work.

1. What is the total increase in monthly income?

2. What is the total increase in monthly expenses?

3. Which is greater, the increase in expenses or the increase in income?

4. Cable TV costs $30 a month. You were only paying half this amount when you had a roommate. How much more do you pay now for cable TV each month?

5. You cancel cable TV. What happens to your monthly expenses?

6. **CRITICAL THINKING** You increase your hours to 8 hours a week at Don's Ice Cream Shop. Your pay is $6 an hour. How much do you make in a month? (Hint: There are 4 weeks in a month.)

Maintaining Skills

Estimate each sum or difference. Round to the nearest dollar. Then add or subtract.

1. $25.42 + 36.18	2. $42.58 − 9.49	3. $27.67 + 3.45	4. $76.52 − 63.14

Compute.

5. $61.72 × $2.50

6. $23.18 + $4.89

7. $6.11 − $2.26

8. $47.00 − $12.24

9. $1.36 × 7.25

10. $48.27 + $3.98

2-6 ▶ What Happens in Emergencies?

What happens to your budget when emergencies come up? This is your unlucky month. Your computer is broken, and it must be repaired. The electronics store said it will cost $185 to fix it.

Soon you realize your wallet is gone. You not only lose your favorite wallet, but the $45 inside it too.

You think things can't get any worse until you sprain your wrist playing basketball. You go to the emergency room. The doctor takes X-rays and puts your wrist in a splint. The bill is $432. It isn't covered by your health insurance plan.

▶ **EXAMPLE 1**

How much extra money do you need this month to cover these emergencies?

Add the unexpected expenses.		
	$185	computer repair
	45	lost money
+	432	doctor bills
	$662	emergency expenses

You need an extra $662 this month.

Things get worse. For five days you can't work. You have already used your sick time, so you lose a week's pay, which is $325. You also lose a week's pay at Don's Ice Cream Shop. Your lost pay at Don's totals $42.

▶ **EXAMPLE 2**

What is the total of your emergency expenses and lost income?

Add your emergency expenses and your lost income.		
	$662	emergency expenses
	325	lost pay–regular job
+	42	lost pay–Don's
	$1,029	total

The total of emergency expenses and lost income is $1,029.

Wordwise
When a person owes money they are said to be *in debt*. When the debts are paid off, that person is *out of debt*.

To cover emergency expenses and lost income, you need $1,029. You only have $300 in your savings. You need to find a way to make up the lost income and pay your **debt**. Debt is money you owe. You contact the hospital and electronics shop to see if you can pay off your debt monthly.

You decide to transfer $30 from clothing and $40 from recreation into a monthly repayment fund. This fund becomes a fixed expense in your budget.

▶ **EXAMPLE 3**

How many months will it take you to get out of debt?

STEP 1 Subtract your savings to find the total debt.

$$\begin{array}{r} \$1,029 \\ -\quad 300 \\ \hline \$729 \end{array}$$ emergency expenses
savings

Here's a Tip!
When you divide, you may get a quotient such as 3.5 people or 10.4 months. You cannot have a half person, and bills are paid only once a month. In these cases, round up to the nearest whole number.

STEP 2 Add the money you will transfer to find the monthly repayment.

$$\begin{array}{r} \$30 \\ +\quad 40 \\ \hline \$70 \end{array}$$ from clothing
from recreation

STEP 3 Divide the debt by the monthly repayment.

$729 ÷ $70 ≈ 10.4 months

You will be out of debt in about 11 months.

Practice and Apply

Use the information above to answer each question.

1. You cancel cable TV, which costs $30 a month. You add this money to the monthly repayment fund. Now how many months will it take to pay off your debts?

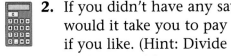

2. If you didn't have any savings, how many months would it take you to pay off your debts? Use a calculator if you like. (Hint: Divide the total debt by the amount of the monthly repayment.)

3. **IN YOUR WORLD** Write about a time when you had an emergency. Did it cost you money? If so, how much?

2·7 ▶ Focus: Finding an Average

How can you plan for an expense if the expense varies? You can find the **average** over a certain amount of time.

▶ **EXAMPLE**

Here's a Tip!
You can save money if you make your own lunch and bring it to work or school.

Alice wrote down her lunch expenses for five days. What was her average lunch expense?

Monday	$3.50
Tuesday	$4.50
Wednesday	$3.35
Thursday	$5.50
Friday	$4.25

STEP 1 Add the expenses.

$3.50 + $4.50 + $3.35 + $5.50 + $4.25 = $21.10

STEP 2 Divide by the number of expenses.

$21.10 ÷ 5 = $4.22

Alice's average lunch expense was $4.22.

Skills Practice

 Find each average. Use a calculator if you like.

1. $6, $18, $16, $20

2. $3, $7, $10, $18

3. $10, $25, $30, $20

4. $8.00, $19.10, $14.00

5. $15.20, $17.50, $12.90

6. $2.50, $3.85, $4.30, $3.35

Everyday Problem Solving

Use Linda's grocery receipt to solve each problem.

1. What was Linda's average expense for the 4 food items? For the 3 personal care items?

2. What was Linda's total average expense for all the items?

```
      Shop-n-Go
Apples..........$3.69
Soup...........$1.59
Crackers........$2.99
Juice..........$3.85
Toothpaste......$4.19
Deodorant.......$2.89
Dental Floss....$2.22

Total.........$21.42
```

How Do You Budget Extra Money?

Imagine that you have no debts and you are getting a promotion at the factory. Congratulations! Your monthly pay at the factory will be $250 more.

Wordwise
A *promotion* is an advance in your job. It means you are given a higher position with more responsibility, and often more money.

Your friends have lots of suggestions. Buy a new snowboard! Go on a vacation! But your first priority is to quit your second job at Don's Ice Cream Shop. This will subtract $168 from your original monthly income of $1,468.

▶ **EXAMPLE**

What is your new monthly income?

STEP 1 Subtract your pay at Don's from your original monthly income.

$1,468	original income
− 168	pay from Don's
$1,300	income without raise

STEP 2 Add your raise to find your new monthly income.

$1,300	income without raise
+ 250	raise
$1,550	new monthly income

Your new monthly income is $1,550.

Practice and Apply

Use the example above. Solve each problem. Show your work.

1. How much more is your new monthly income than your original monthly income?

2. Now you make $1,550 a month. If there are 4 weeks in a month, how much do you make each week?

3. If you kept your job at Don's Ice Cream Shop, how much would your total income be each month?

4. CRITICAL THINKING Your new monthly income is $1,550. Don asks you to work one evening a week. You will make $19.50 a week at Don's. How much will you make altogether each month?

How Do You Spend Your Extra Money?

You were promoted, so you have some extra money each month. Now you must decide how to spend it.

You decide to buy a used car. You now must also pay for gas and car insurance. However, an inexpensive laundromat just opened in town. This will reduce your monthly laundry expenses. You will save for gifts and emergencies, and you decide to budget more for recreation and clothing. But most of your raise goes to paying for the car.

Here is your new budget.

Your New Monthly Budget	
Monthly Income	$1,550
Fixed Expenses	
Rent	$483
Car insurance	$110
Car loan	$199
Variable Expenses	
Gasoline	$40
Utilities	$156
Phone	$30
Laundry ($4 a week × 4 weeks)	$16
Clothing	$85
Groceries and personal care	$206
Recreation ($45 a week × 4 weeks)	$180
Savings	
Savings for gifts and emergencies	$45
Total	$1,550

Your new monthly budget allows you more money to do the things you enjoy. Not only can you spend more money on recreation now, but you have a car to get to places. This makes life a lot easier.

Practice and Apply

Use your new monthly budget on page 36. Solve each problem. Show your work.

1. What are the total monthly expenses for your new car?

 2. The price of the car you want is $5,582. You put $2,000 down. How many months will you need to pay the car loan? Use a calculator if you like.

3. After 5 months, your car needs a new battery and 2 new tires. These will cost $180. Will you have saved enough money at this time for these items? Explain.

4. You drive a friend to work each day. She pays you $5 a week for gas. You put that money into your budget for clothing. Now how much do you have for clothing each month?

5. Suppose you make some long-distance phone calls. Your phone bill is $48.50. How much did you budget for the phone bill? How much will you need to transfer from other variable expenses to pay for the extra charges?

6. **IN YOUR WORLD** Suppose you have $140 a month to spend on recreation. What activities will you choose? How much will each activity cost you?

Maintaining Skills

Multiply or divide.

1. $14.98
    $\times$ 12

2. $4\overline{)\$116.36}$

3. $72.98
     $\times$ 6

4. $5\overline{)\$220.55}$

5. $183.72 ÷ 3

6. $288.72 ÷ 6

7. $29.48 × 11

8. $15.42 × 6

9. $879.48 ÷ 7

10. $18.02 × 13

11. $963.27 ÷ 9

12. $76.45 × 4

Solve each problem. Show your work.

1. Ryan budgets $35 a week for lunch for 5 days. What is his average lunch expense each day?

2. Joe pays $25 a month to have his uniforms cleaned. How much does he pay in a year?

3. Jen spent $11.00, $12.50, $14.00, and $10.50 on gas for her car in the last four weeks. What was her average weekly expense for gas?

4. Lewis wants to buy groceries that cost $3.46, $2.42, $5.16, $4.87, $6.04, $1.79, $2.29, $3.15, $3.89, and $2.76. He has $35.00. Estimate to see if he has enough money. Round up to the nearest dollar.

5. **OPEN ENDED** Valerie budgeted $85 a month for recreation. Going to the movies costs $9, renting a canoe costs $25, bowling costs $12, horseback riding costs $60, and a concert ticket costs $30. Which activities can Valerie afford in one month?

Calculator

You should always check the answer on a calculator display to see if it is reasonable.

Jane used a calculator to add $14.25, $1.39, and $5.89. The calculator display showed 159.14 . This answer is not reasonable. Jane keyed in $139 instead of $1.39.

Decide if each answer is reasonable. Write *Yes* or *No*. If the answer is not reasonable, find the correct answer and the error that was probably made.

1. $28.62 + $39.50 + $17.89   34359

2. $350.20 − $70.00   280.20

3. $890 × 12   106.80

4. $52,000 ÷ 52   1000

ON-THE-JOB MATH:
Administrative Assistant

Lindsay is an administrative assistant in a small company. She takes care of customer service. She writes letters, organizes and updates files, and keeps track of employee attendance. She also orders supplies and balances the service and supply budget.

This month, all employees are getting new electric pencil sharpeners. This will add $150 more to the cost of office supplies this month. Lindsay must balance the budget. She will need to transfer some money.

Office Service and Supply Budget for December: $1,500			
Fixed Expense	Cost	Variable Expense	Cost
Photocopy equipment and service	$180	Office supplies	$200
Cable modem and Internet	$250	Facility supplies (toilet paper, light bulbs, etc.)	$140
Mail and packaging costs	$230	Snow plow fees for plowing lot	$100
Pantry supplies (coffee, cups, etc.)	$200	Employee celebrations	$200

Use the information above to solve each problem.

1. How much does Lindsay need to transfer to pay for the pencil sharpeners?

2. Can Lindsay transfer enough money from photocopy equipment and service to pay for the pencil sharpeners? Why or why not?

3. Which variable expenses may not occur every month?

You Decide

Would you transfer money from fixed expenses or variable expenses to buy the pencil sharpeners? List the budget items you would transfer money from.

Summary

You can keep track of how you spend your money by preparing a budget.

You must find ways to make your income meet your needs and decide on the best way to spend your money.

To balance a budget you can cut back on spending, take out a loan, increase your income by getting another job, or eliminate certain variable expenses.

Two ways you can cope with financial emergencies are transferring money from one budget item to another and beginning a repayment fund.

When you have extra money you must include it in your budget. You can choose to spend it or save it.

Estimating helps you as a consumer. When estimating expenses it is usually better to round up. When estimating your income, it is usually better to round down.

Finding an average can help you plan your budget.

balanced budget

budget

debt

estimate

predict

transfer

Vocabulary Review

Complete the sentences with the words from the box.

1. Money that is owed is called _____.

2. To _____ is to make a guess about the future.

3. To _____ is to move money from one budget line to another.

4. The answer you get when you use rounded numbers is an _____.

5. A _____ is a plan where income is equal to the sum of expenses and savings.

6. A plan for spending and saving money is called a _____.

Chapter Quiz

Use Kurt's monthly budget to solve each problem.

Kurt's Monthly Budget

Monthly Income _____ $1,800

Expenses

Rent	$450	Groceries	$86
Car loan	$159	Personal care	$24
Car insurance	$90	Recreation	$42
Utilities: gas	$50	Clothing	$46
Utilities: electric	$63	Gasoline	$34

Savings _____ $450

1. What is the total amount Kurt can spend in one month?

2. How much does Kurt pay a month for utilities?

3. Kurt owes $2,000 on his car loan. How long will it take to pay this?

4. On average, how much does Kurt spend weekly on recreation?

5. An emergency occurs and Kurt needs an additional $82 a month. From which budget items should he transfer money?

Maintaining Skills

Compute.

1. $135.67
  + 295.43

2. $1,312
  − 584

3. $317.84
  × 42

4. 3)$183.42

Estimate the sum or difference. Round to the nearest dollar.

5. $52.00 + $19.82 + $29.08 + $114.20   6. $68.84 − $13.29

Unit 1 **Review**

Use the circle graph to answer each question. Write the letter of the correct answer.

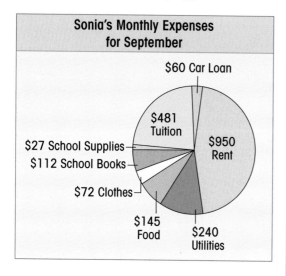

Sonia's Monthly Expenses for September

- $60 Car Loan
- $481 Tuition
- $950 Rent
- $27 School Supplies
- $112 School Books
- $72 Clothes
- $145 Food
- $240 Utilities

1. What were Sonia's total monthly expenses for September?

- **A.** $2,087
- **B.** $992
- **C.** $1,137
- **D.** $2,078

2. Suppose Sonia had an emergency in September. From which category could she have transferred money in her budget?

- **A.** Rent
- **B.** Clothes
- **C.** Car loan
- **D.** Food

3. What were Sonia's total expenses for school?

- **A.** $620
- **B.** $593
- **C.** $139
- **D.** $508

4. How much of Sonia's expenses were variable?

- **A.** $1,491
- **B.** $596
- **C.** $356
- **D.** $60

5. How much would Sonia spend on her rent and car loan in a year?

- **A.** $6,060
- **B.** $10,680
- **C.** $12,120
- **D.** $12,264

6. On average, how much did Sonia spend a week on food?

- **A.** $36.25
- **B.** $72.50
- **C.** $145
- **D.** $50

Challenge

Sonia's yearly income is $27,000. How much was Sonia able to save in September?

Chapter 3 **Your Salary**

Chapter 4 **Your Take-home Pay**

When you work, you earn a paycheck. You can use your monthly pay for expenses like rent and food.

The circle graph shows how Sarah spends her monthly pay on expenses. Use the graph to answer each question.

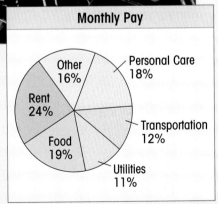

Monthly Pay

Other 16%

Personal Care 18%

Rent 24%

Food 19%

Transportation 12%

Utilities 11%

1. Which monthly expense uses the greatest percent of pay?

2. What percent of monthly pay is for food?

3. What is the total percent of pay used for rent, transportation, and utilities? (Hint: Add the three percents to find the total.)

43

The money a waitress earns may include tips. Should a waitress keep a record of her tips? Explain.

Learning Objectives

LIFE SKILLS

- Determine yearly, monthly, and weekly salaries.
- Read a time card and find overtime pay.
- Find income based on salary and tips.
- Determine job-related expenses.
- Determine commission and find income based on salary and commission.

MATH SKILLS

- Add, subtract, multiply, and divide with money.
- Find the percent of a number.
- Round money amounts to the nearest cent.

Words to Know

salary	the fixed pay for regular work
yearly salary	the amount of money earned in one year or 12 months
employee	a person who is hired to work for another person or business
monthly salary	the amount of money earned in one month
weekly salary	the amount of money earned in one week or 7 days
employer	a person or business that gives work to another person
time card	a record of the hours an employee works
overtime	the hours worked beyond the number of hours in the regular work week
hourly wage	the amount of money paid for one hour of work
percent	a part of a whole, based on 100 parts
commission	a fee that is sometimes paid for sales work; a commission is usually a percent of the total sale

Project: Finding Work

You've decided to change jobs. Look in the newspaper or on the Internet to find three possible jobs. A useful Web site is www.usajobs.opm.gov/. This is the official site for U.S. government jobs. Decide which job is best. Consider the following.

- What is the yearly salary or hourly wage?
- How many hours will you work a week?
- Is the job close to home?
- Will you need any new skills?
- Do you need more education?

How Do You Calculate a Yearly Salary?

People change jobs for many reasons. If you need a new job, how could you find one?

- Put in an application for a different job where you already work.

- If you are in school, talk to your school guidance counselor or career center.

- Ask friends if they know of any job openings.

- Go to an employment agency.

- Look at employment ads in a newspaper.

Here is an ad from a newspaper.

Counter Person for Scotts Valley Supermarket. Seeking courteous & reliable applicant. Must be at least 18 years old. $7.25/hr. Apply in person. 582 Mt. Hermon Rd. 889-555-5578

Wordwise
The symbol / reads *per.*
Per means *each.*

When finding a new job, it is important to consider the **salary.** If you know the hourly pay for a job, you can find the **yearly salary.** You need to know the number of hours the **employee** must work each week.

EXAMPLE

You call the number in the ad above and learn that a counter person works 40 hours a week. What is the yearly salary? (Hint: There are 52 weeks in a year.)

STEP 1   Multiply the hourly pay by the number of hours worked in a week.
$7.25 × 40 = $290 per week

STEP 2   Multiply the weekly pay by 52.
$290 × 52 = $15,080 per year

The yearly salary is $15,080.

Practice and Apply

Here are two employment ads from a newspaper. Use them to solve each problem. Show your work.

Driver needed for hauling service. Must have phone & current driver's license. $14/hr. to start. $18/hr. after 28 days. Call Jerry 777-555-7314

Nurse needed for long-term care facility. RN $21.50/hr. LPN $14/hr. Paid Bonus: 1 wk. vacation after 6 mos. + medical benefits. Call Peggy 777-555-9822

1. How much per hour will the driver be paid during the first 28 days? What will the driver be paid for a 40-hour work week during this time?

2. What will the driver be paid for a 40-hour work week after 28 days?

3. A driver works 40 hours a week for the first two months on the job. How much more per week does the driver make after 28 days?

4. A driver works 40 hours a week for the first year on the job. How much does the driver earn for the year? (Hint: The driver earns $14 an hour for the first 4 weeks he works. Then he earns $18 an hour.)

5. RN means *registered nurse*. LPN means *licensed practical nurse*. What is the hourly pay for the LPN? How much more is the RN paid for a 40-hour work week than the LPN?

6. What is the yearly salary for the RN working 40 hours a week? How much more is an RN paid in a year than an LPN?

7. **CRITICAL THINKING** What else besides the hourly wage is important to you when choosing a job? If you are married with children, does that affect your decision when choosing a job? Explain your thinking.

The employment agency tells you about a job that pays $12,000 per year. How can you find out how much you would earn per month and per week?

Here's a Tip!
Sometimes you may need to round to the nearest cent. Look at the third digit after the decimal point. If it's 5 or higher, round up. If it's 4 or lower, round down.
$230.76923 → $230.77
$145.583 → $145.58

To find a **monthly salary**, divide the yearly salary by 12 because there are 12 months in a year.

$12,000 ÷ 12 = $1,000 per month

To find a **weekly salary**, divide the yearly salary by 52 because there are 52 weeks in a year.

$12,000 ÷ 52 = $230.76923

$230.76923 is about $230.77 per week.

Practice and Apply

Find the monthly and weekly salary for each job. Use a calculator if you like. Round to the nearest cent.

	Job	Yearly Salary	Monthly Salary	Weekly Salary
1.	Bank teller	$14,367	?	?
2.	Painter	$31,886	?	?
3.	Dental assistant	$51,334	?	?
4.	Travel agent	$25,000	?	?
5.	Sheet metal worker	$32,010	?	?
6.	Computer technician	$36,460	?	?
7.	Police officer	$39,790	?	?
8.	Bus driver	$30,368	?	?

Use the salaries you found on page 48 to solve each problem.
Show your work.

9. A police officer marries a bank teller. What is their combined income each month?

10. A bus driver marries a dental assistant. What is their combined income each week?

11. How much can a computer technician earn in two years? Assume there are no raises in that time.

12. Who earns more per year, a computer technician or a sheet metal worker? How much more?

13. Who earns more per year, a painter or a police officer? How much more?

14. **IN YOUR WORLD** Which job on the list would you most like to have? Why? Which job would you least like to have? Why?

Maintaining Skills

Compute.

1. $405.75 + 395.25	2. $1,200 − 869	3. $215.50 × 35	4. $600.29 − 450.77
5. $383.50 − 345.55	6. $300.55 × 8	7. $5,209 + 485	8. $22.40 × 14

9. $4,800 ÷ 12   10. $16.30 ÷ 5   11. 12)$2,535   12. 52)$33,735

Find the average of each set of numbers.
13. $8.25, $102.00, $98.39, $6.00   14. $33, $140, $7, $10, $30

Your **employer** may ask you to use a **time card** to keep track of the number of hours you work. A clerk may write down your hours on the card. Or, at the beginning and end of the workday, you may put your card into a special clock that stamps the time. You also need to keep track of any breaks you take during the day.

This time card records how much time Mike spent on his job one day.

Wordwise

A.M. stands for the hours from midnight to noon. P.M. stands for the hours from noon to midnight. 12 A.M. is midnight and 12 P.M. is noon.

Name: Mike Quinn			Job Title: Assembler		
Hourly Rate: $7			Week Ending: 6/16		
Day	**Date**	**Time In**	**Time Out**	**Break**	**Total**
Monday	6/12	4 P.M.	12 midnight	1 hour	?

▶ **EXAMPLE 1**

How many hours did Mike work on Monday?

STEP 1 Subtract to find how many hours are between 4 P.M. and midnight.

$$\begin{array}{r} 12 \text{ midnight} \\ - \quad 4 \text{ P.M.} \\ \hline 8 \text{ hours} \end{array}$$ work ended
work began

STEP 2 Subtract the break.

$$\begin{array}{r} 8 \text{ hours} \\ - \quad 1 \text{ hour} \\ \hline 7 \text{ hours} \end{array}$$

Mike worked 7 hours on Monday.

Sometimes you need to find the number of hours in a time period that goes beyond noon or midnight. It is easier to think about this as two separate time periods. First find the number of hours in each time period. Then add the number of hours in each time period to find the total.

▶ **EXAMPLE 2**

How long is the time between 7 A.M. and 4 P.M.?

STEP 1 Subtract to find how many hours are between noon and 4 P.M.

$$\begin{array}{r} 4 \text{ P.M.} \\ - \underline{12 \text{ noon}} \\ 4 \text{ hours} \end{array}$$

STEP 2 Subtract to find how many hours are between 7 A.M. and noon.

$$\begin{array}{r} 12 \text{ noon} \\ - \underline{7 \text{ A.M.}} \\ 5 \text{ hours} \end{array}$$

STEP 3 Add the result to the number of hours between noon and 4 P.M.

5 hours from 7 A.M. to noon
+ 4 hours from noon to 4 P.M.
9 hours

There are 9 hours between 7 A.M. and 4 P.M.

Practice and Apply

Find the total hours Tara worked each day. Use that information to solve problems 6–8.

	Day	Date	Time In	Time Out	Break	Total Hours
1.	Monday	6/12	5 P.M.	12 A.M.	1 hour	?
2.	Tuesday	6/13	7 A.M.	5 P.M.	1 hour	?
3.	Wednesday	6/14	12 A.M.	7 A.M.	1 hour	?
4.	Thursday	6/15	5 P.M.	12 A.M.	1 hour	?
5.	Friday	6/16	8 A.M.	5 P.M.	1 hour	?

6. How many hours did Tara work this week?

7. On average, how long did Tara work each day?

8. **CRITICAL THINKING** Suppose that Tara worked 4 hours on Saturday. On average, how long did Tara work each day?

How Do You Calculate Overtime?

Sandra is excited about working **overtime**. She can earn extra money by working more than the regular number of hours she is scheduled to work.

For some jobs, the **hourly wage** for overtime is more than the regular hourly wage. Overtime may be paid as time and a half or even double time.

► **EXAMPLE**

Last week Sandra worked 45 hours. She gets paid overtime for any hours over 40 hours per week. Her regular hourly wage is $6. Overtime is paid as time and a half. How much did Sandra earn for the hours she worked overtime? How much did she earn for the week?

Here's a Tip!
Time and a half means $1\frac{1}{2}$ times. Multiply by $1\frac{1}{2}$ or 1.5.

STEP 1   Find the overtime hourly wage. Multiply the regular hourly wage by the overtime rate.

$6 × 1.5 = $9

STEP 2   Subtract to find the number of overtime hours worked.

45 − 40 = 5 hours

STEP 3   Multiply the overtime hourly wage by the number of overtime hours worked.

$9 × 5 = $45

Sandra earned $45 for the hours she worked overtime.

STEP 4   Multiply to find the regular pay for 40 hours.

$6 × 40 = $240

STEP 5   Add the overtime pay to the regular pay to find the total weekly pay.

$240 + $45 = $285

Sandra earned $285 for the week.

Practice and Apply

Solve each problem. Show your work.

1. Ariel is working at a meat-packing plant 5 nights a week. Her regular wage is $11 an hour. She earns time and a half for any overtime hours. This week she worked 9 hours of overtime. How much will Ariel earn for overtime this week?

2. Isaac is working as a security guard. He earns $8.45 an hour. He earns double time for any overtime over 40 hours. Last week he worked 44 hours. How much did he earn last week? (Hint: Double time means 2 times the regular hourly rate.)

3. Michelle's regular hourly wage at the candy factory is $9.86 for a 35-hour work week. She is paid double time for working on a holiday. Last week she worked 10 hours overtime on a holiday. How much more did she earn last week than she earns in a 35-hour work week?

4. At the Conrad Bakery plant, 200 employees each worked 5 hours overtime this week. Each employee has a regular wage of $9.40 an hour. Overtime is paid at a rate of time and a half. How much in overtime will the plant pay this week? Use a calculator if you like.

5. **WRITE ABOUT IT** You are paid $8.90 an hour for a 35-hour work week. You would like to work 5 hours a week overtime to earn extra money. Your company pays time and a half for overtime. Write a memo to your boss, asking that you be allowed to work overtime.

How Do You Calculate Your Income When You Make Tips?

Wordwise
A skycap is an employee who carries passenger baggage at an airport. A bellhop does the same for people staying at a hotel.

If you enjoy meeting people, you might enjoy a job as a waiter, taxi driver, hairstylist, hotel maid, skycap, or bellhop.

People in these jobs sometimes get tips as well as a salary. When you calculate income, include tips.

▶ **EXAMPLE**

Sam works 40 hours a week as a bellhop. He earns $9.70 an hour. Last week he also received $375 in tips. What was Sam's income last week?

STEP 1   Calculate the weekly salary. $9.70 × 40 = $388

STEP 2   Add the tips. $388 + $375 = $763

Sam's income was $763 last week.

Practice and Apply

Find the weekly income for each employee.

	Employee	Hourly Wage	Hours Worked	Weekly Tips	Weekly Income
1.	Todd (waiter)	$6.42	30	$450	?
2.	Sally (taxi driver)	$8.34	40	$200	?
3.	Joe (café musician)	$7.00	15	$75	?
4.	Cindy (hotel maid)	$8.46	20	$50	?

5. **CRITICAL THINKING** Each employee in the chart above works 5 days a week. Choose one employee. Calculate the average amount of money that person earned from tips in one day.

What About Expenses on the Job?

Sam enjoys his job as a bellhop. He likes helping people with their luggage at the hotel. He wears a uniform so people will know that he is a bellhop. Sam must buy or rent uniforms and have them cleaned regularly. These job-related expenses mean that Sam has less money for his other needs and wants.

▶ **EXAMPLE**

Sam spent $340 to buy uniforms this year. Dry cleaning costs him about $320 per year. How much will Sam spend on those expenses this year?

Add to find the total cost.

$340 + $320 = $660

Sam will spend $660 for those expenses this year.

Practice and Apply

Solve each problem. Show your work.

1. Bob works as a carpenter and needs to buy some tools for work. A framing hammer costs $22. A 4-foot level costs $18. A carpenter's square costs $7. What are Bob's total expenses for these tools?

2. Cary works 5 days a week in a nursing home. Her weekly paycheck is $786. She spends $5.25 to dry-clean her uniform each week. She also spends $2.25 on a roundtrip bus ticket to get to work each day. How much money from her paycheck does Cary have left after paying these expenses?

3. **WRITE ABOUT IT** Look at the employment section in the local newspaper. Choose a job that you find interesting. List the job-related expenses you might have during your first week on the job.

3·7 Focus: Finding the Percent of a Number

Erin is a waitress. She must calculate the total cost of each food order. The total includes a sales tax.

Sales tax is a **percent** of the cost of an item. To calculate with a percent, first write the percent in decimal form. Drop the percent sign and move the decimal point two spaces to the left. Include zeroes if needed.

$$5\% = 5.\% \rightarrow 0.05 \qquad\qquad 3.2\% \rightarrow 0.032$$

Here's a Tip!
5% means 5 out of 100, or 5 hundredths. The decimal 0.05 is read as 5 hundredths.

To find the percent of a number, multiply the number by the decimal form of the percent.

> **EXAMPLE 1**

Erin calculates a 6% sales tax on the total cost of a food order. What is the sales tax on a $30 food order?

STEP 1 Change the percent to a decimal.
$6\% = 0.06$

STEP 2 Multiply the dollar amount by the decimal.
$\$30 \times 0.06 = \$1.80$

The sales tax is $1.80.

If the service is good, a customer leaves a tip.

> **EXAMPLE 2**

Mr. Books took his family out to eat. The total for the food order came to $72. The sales tax is 8%. He gave the waiter a 15% tip. How much did Mr. Books pay for the meal?

STEP 1 Find the amount of the sales tax.
8% of $\$72 \longrightarrow 0.08 \times \$72 = \$5.76$

STEP 2 Find the amount of the tip.
15% of $\$72 \longrightarrow 0.15 \times \$72 = \$10.80$

Did You Know?
You calculate the tip on the total cost of the food order, not including sales tax.

STEP 3 Add to find the total for the meal.
$\$72 + \$5.76 + \$10.80 = \$88.56$

Mr. Books paid $88.56 for the meal.

Skills Practice

Write each percent as a decimal.

1. 24%

2. 20%

3. 55%

4. 1%

5. 4.7%

6. 6.99%

7. 7.05%

8. 10.05%

9. 100%

Find the percent of each number.

10. 20% of 50

11. 70% of 10

12. 14% of $350

13. 5% of $25

14. 35% of $250

15. 40% of $150.50

16. 7.5% of $46

17. 1.5% of $8

18. 100% of $78

Everyday Problem Solving

The menu shows the price of a complete dinner, including a salad, dessert, and coffee or tea. Use it to solve each problem. Show your work.

Today's Menu

Main Course	Price
Shrimp	$18.50
Salmon Steak	$19.75
Pork Roast	$20.95
Chicken Cutlets	$17.25
Beef Stew	$16.95
Vegetable Lasagna	$16.50
Lamb Chops	$21.45

1. Mary served a table of three people who ordered beef stew, lamb chops, and pork roast. How much was their total bill? Include a 3% sales tax.

2. Mary gave the Jackson family extra service because they had small children. Their order totaled $50.70 without sales tax. They left a 20% tip for Mary. How much was the tip?

3. Three people at one table all ordered shrimp dinners. Three people at a second table ordered salmon, chicken, and vegetable lasagna. Which group paid more tax? Explain.

3-8 ▶ What Is a Commission?

Many jobs require selling goods or services. In some of these jobs, your income may depend on how much you sell. When your income is based on sales, you are earning a **commission**.

▶ **EXAMPLE**

Alice receives a 2.5% commission on every piece of furniture she sells. One day she sold a table for $450. What was her commission for that sale?

STEP 1 Change the percent to a decimal.
2.5% = 0.025

STEP 2 Multiply the amount of the sale by the decimal.
$450 × 0.025 = $11.25

Alice received $11.25 as a commission for that sale.

Practice and Apply

Find each commission.

	Salesperson	Amount of Sales	Rate of Commission	Commission
1.	Susan	$124,000	3%	?
2.	Marty	$3,887	4%	?
3.	Stephen	$146	6.5%	?
4.	Denise	$53,990	2.1%	?
5.	Carlos	$12,855	8%	?

6. CRITICAL THINKING Angelica made sales of $5,000. She receives a 3% commission. How much would her commission be if her sales doubled?

3-9 What If You Earn a Salary Plus Commission?

Sometimes you can earn a salary plus commission.

EXAMPLE

Jack earns $144 a week plus a 2% commission selling electronics part-time. Last week his sales were $1,850. How much did Jack earn last week?

STEP 1 Change the percent to a decimal.
2% = 0.02

STEP 2 Multiply the amount sold by the decimal to find the commission.
$1,850 × 0.02 = $37

STEP 3 Add the commission to the salary.
$144 + $37 = $181

Jack earned $181 last week.

Practice and Apply

Find each commission and total income. Use a calculator if you like.

	Weekly Sales	Rate of Commission	Commission	Weekly Salary	Total Income
1.	$6,400	3%	?	$540	?
2.	$26,887	2%	?	$350	?
3.	$1,450	1%	?	$445	?
4.	$5,000	2.5%	?	$260	?
5.	$1,500	3%	?	$560	?
6.	$8,400	4.5%	?	$620	?

7. WRITE ABOUT IT Jack earns a commission and a salary selling electronics. How does this benefit Jack? How does it benefit his employer? Does it benefit you as a consumer? Explain.

Solve each problem. Show your work.

1. Leroy is a house painter who has just been paid for a job. He received $960 for the job. The paint cost $85. How much money did Leroy make after paying expenses?

2. Belinda sells furniture at the Fantasy Furniture Company. She earns $430 plus 2% commission on her sales each week. Last week her sales were $12,470. How much did Belinda earn last week?

3. Celia is the payroll clerk at the Spinaca Cannery. This week, 6 workers each put in 5 hours of overtime. They earn $9.50 an hour. Overtime is paid at double that rate. How much money does Celia include in the payroll for overtime?

4. **OPEN ENDED** Tom received $250 in tips, working as a waiter for 5 days. How much could he have gotten in tips each of the 5 days he worked?

Calculator

On some calculators, you can use the percent key to find the percent of a number. The calculator changes the percent to a decimal.

Find 50% of 25.

Press: [2] [5] [×] [5] [0] [%] ⟨ *12.5* ⟩

Find the value of each expression.

1. 50% of 75

2. 15.5% of 10

3. 23% of $200

4. 8.2% of $480

5. 75% of 5,300

6. 12% of $30

7. 10% of 4.5

8. 200% of $100

DECISION MAKING:
Would You Choose This Job?

Kim is working as a teacher's aide. She enjoys children, but she also enjoys the world of nature. Kim is thinking about changing jobs. She sees an ad for an assistant at a nature education center. Kim calls to ask about the job.

What Kim Learns About the Job

- The job includes teaching children about the natural world.
- The assistant will use the computer to keep records and prepare class materials.
- The job often requires being outdoors, leading a class, or working on projects such as labeling plants along the trails.
- The work schedule is four days a week, Wednesday through Saturday.

Answer each question. Use the chart above.

1. Kim is interested in the job and schedules an interview. What skills does she already have that she could talk about in the interview?

2. Kim is not experienced with the computer software that the nature center staff uses. If you were Kim, how would you handle this concern in an interview?

3. How might this job give Kim a chance to learn new skills?

4. What advantages might this job offer Kim as compared to her present job? What disadvantages might it have?

You Decide

The salary for this job is less than what Kim is earning now. However, the job does not require Kim to work more than four days a week. If Kim is offered the job, how might this affect her decision?

Summary

You can calculate your yearly and monthly salary if you know the number of hours you work each week and the hourly wage you earn.
A time card is a record of the hours worked each day.
To find overtime pay, multiply the number of overtime hours by the overtime rate.
Sometimes income will include salary plus money earned as tips.
Job-related expenses reduce the money you have for other expenses.
A salesperson may be paid a commission as well as a salary. To find the commission, multiply the amount of sales by the percent commission.

commission

employee

employer

hourly wage

monthly salary

overtime

percent

time card

weekly salary

yearly salary

Vocabulary Review

Complete each sentence with a word from the box.

1. A person or business that gives work to another person is an _____.

2. A _____ is a fee that is sometimes paid for sales work.

3. A record of the hours an employee works is called a _____.

4. The hours worked beyond the number of hours in the regular work week is _____.

5. A person who is hired to work for another person or business is an _____.

6. A _____ is the amount of money earned in one year.

7. An _____ is the amount of money paid for one hour of work.

8. A _____ describes part of a whole, based on 100 parts.

9. The amount of money earned in one month is a _____.

10. The amount of money earned in one week is a _____.

Chapter Quiz

Solve each problem. Show your work.

1. Velma saw a job advertised in the paper. It pays $8 an hour for a 40-hour work week. How much could Velma make in four weeks?

2. Ralph is working as a stock person for $6.25 an hour. He works 35 hours a week. What is his yearly salary?

3. Gabe's yearly salary as a bus driver is $21,000. What is his monthly salary?

4. Magda started work at 9 A.M. and stopped at 2 P.M. How many hours did she work?

5. Indra worked overtime for 8 hours last week. Her regular hourly wage is $7.50. Her overtime rate is 1.5 times her regular wage. How much did she earn in overtime?

6. Nguyen is a waiter. One night he earned $40 in tips. His hourly wage is $7. He worked from 6 P.M. until 11 P.M. How much did he earn that night?

7. Rosemary earns a 3.5% commission on furniture she sells. One day she sold a sofa for $960. How much was her commission?

Maintaining Skills

Compute.

1. 45% of 35

2. 3.5% of $250

3. 25% of 445

4. 10.5% of 50

5.
$$\begin{array}{r} \$500.00 \\ - 458.88 \end{array}$$

6.
$$\begin{array}{r} \$605 \\ \times 4.5 \end{array}$$

7. $12\overline{)\$31,200}$

8.
$$\begin{array}{r} \$227 \\ + 459 \end{array}$$

The money a doctor earns will be greater than her take-home pay. Take-home pay is money you earn less deductions such as income tax. Where does income tax go?

Learning Objectives

LIFE SKILLS

- Identify earnings deductions and determine net pay.
- Interpret an earnings statement.
- Determine taxable income and use a tax table.
- Calculate an income tax refund or balance due.
- Compare two health insurance options.
- Identify savings options and retirement benefits.

MATH SKILLS

- Interpret a circle graph showing percents.
- Find percent, given the whole and the part.

Words to Know

gross pay	amount of money earned by an employee
net pay	amount of money received after deductions have been taken
deduction	money taken from gross pay for things such as taxes, health insurance, or union dues
withheld	kept out; for example, money withheld from a salary
tax	money paid to the local, state, or federal government
exemption	reduction in your income tax for each person you support, including yourself
earnings statement	the paycheck stub listing gross pay, deductions, and net pay
interest	the money a bank pays you for keeping your money there
W-2 form	a federal tax form that shows your income for a year
credit union	a savings and loan business that offers services to company employees or other groups
savings bond	a way to save money; the purchase price of a savings bond is repaid with interest when the bond is sold
401(k) plan	a type of retirement plan into which an employee contributes money from each paycheck; employers may also make a contribution

Project: Understanding Local Taxes

The government uses taxes to pay for things such as public schools, police services, and roads. List five ways that your town uses its tax money. Your town Web site may have this information. How much will your town spend this year for each item on your list?

4·1   What Is a Deduction?

You may be surprised to learn that your take-home pay is less than what you've earned. **Gross pay** is the money you earn on the job. Gross pay is what you earn before deductions. **Net pay** is your take-home pay after deductions. So what is a deduction?

A **deduction** is money **withheld** from your paycheck. Some common deductions are listed below.

Payroll Deductions	
Deduction	**What It's For**
Federal income tax	Money paid to the United States Treasury
Social Security (FICA)	Money paid to a federal plan that helps support older people
Medicare	Money paid to a federal plan that helps pay medical expenses for older people
State income tax	Money paid to your state treasury
Health insurance	Money paid to a plan that helps pay your medical bills
Disability insurance	Money paid to a plan that pays workers if they can't work because of illness or injury
United Way	Money given to charities that help people in need
Union dues	Money paid by members of an organization that helps workers get what they need from their employers
Credit union payments	Money paid for the banking services of a savings and loan business that serves company employees or other groups

Federal and state income **tax**, Social Security, and Medicare are taken from almost every paycheck. However, a few states have no state income tax. In those states, taxpayers still pay a federal income tax.

Federal and state income taxes are not deducted at the same rate for each taxpayer. The more you earn, the higher your taxes can be.

How much you pay in taxes also depends on whether you have an **exemption.** You may have more than one exemption. The number of exemptions you have usually equals the number of people you support. Supporting yourself gives you one exemption. More exemptions mean lower taxes.

FICA and Medicare are deducted as a percent of your salary. The percent can change from time to time.

Wordwise
FICA stands for *Federal Insurance Contribution Act.* It is a form of insurance to protect people against loss of income.

Practice and Apply

Write the letter of the correct answer.

1. What is gross pay?
   A. your full salary before deductions
   B. your salary after deductions
   C. money received for tips

2. What is a deduction?
   A. money added to your salary
   B. money subtracted from your salary
   C. combined net and gross pay

3. What is FICA?
   A. union dues
   B. United Way
   C. Social Security

4. Which deduction is paid to the United States Treasury?
   A. state income tax
   B. federal income tax
   C. United Fund

5. What does an exemption do?
   A. lower your taxes
   B. raise your taxes
   C. give money to someone you support

How Do You Find Net Pay?

Bernie works at a kennel, feeding and bathing animals. He earns $600 every two weeks. But his first paycheck after two weeks was only $432. Where did the rest of his money go?

Here is the **earnings statement** that was attached to Bernie's paycheck. Notice that four deductions were taken from his gross pay.

Name: Bernie Sloan		Social Security Number: 999-99-9999			
Gross Pay	**Federal Income Tax**	**State Income Tax**	**FICA**	**Medicare**	**Net Pay**
$600.00	$90.00	$32.10	$37.20	$8.70	$432.00

The amount of money taken for deductions equals the difference between gross pay and net pay.

▶ **EXAMPLE**

Did Bernie receive the correct net pay?

STEP 1 Add the deductions to find the total.

Here's a Tip!
Line up the decimal points when adding or subtracting money.

$90.00	Federal income tax
32.10	State income tax
37.20	FICA
+ 8.70	Medicare
$168.00	

STEP 2 Subtract the deductions from gross pay.

$600.00	Gross pay
− 168.00	Deductions
$432.00	Net pay

Bernie did receive the correct net pay.

Practice and Apply

 Solve each problem. Show your work. Use a calculator if you like. Use Leroy Johnson's weekly earnings statement for Problems 1–5.

Name: Leroy Johnson		**Social Security Number:** 999-01-0009					
Gross Pay	**Federal Income Tax**	**State Income Tax**	**FICA**	**Medicare**	**Health Insurance**	**United Way**	**Net Pay**
$340.00	$51.00	$7.14	$21.08	$4.93	$28.00	$2.00	$?

1. What is the total amount of deductions taken from Leroy's weekly gross pay?

2. What is Leroy's weekly net pay?

3. Leroy has a paid vacation, and he gets the same paycheck every week. What is Leroy's yearly gross pay?

4. Florida has no state income tax. If Leroy lived in Florida, what would his yearly net pay be?

5. Suppose Leroy is injured on the job and cannot work for a month. Will he receive any money to replace part of his lost earnings that month? Explain.

6. **IN YOUR WORLD** Do you live in a state that has a state income tax? Which states do not have an income tax? (Hint: The Internet can help you answer these questions. Go to www.taxadmin.org)

How Do You Find a Percent of Gross Pay?

Each deduction taken from your gross pay reduces it by a certain percent. A circle graph shows what part of your gross pay is taken for each deduction.

Ariel is working as a dressmaker in a bridal shop. Her gross pay is $400 a week. She pays federal income tax, FICA, and health insurance.

Here's a Tip!
100% means the whole amount. A percent less than 100% means part of the whole. For example, 50% is $\frac{50}{100}$, or $\frac{1}{2}$, of the whole amount.

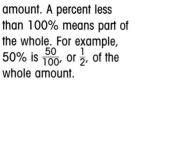

The circle graph shows what percent of Ariel's gross pay is taken for each deduction. Notice that the percents on the circle graph add up to 100%.

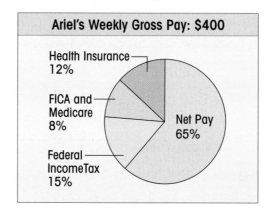

Ariel's Weekly Gross Pay: $400

Health Insurance 12%

FICA and Medicare 8%

Federal IncomeTax 15%

Net Pay 65%

You can use what you know about percents to find how much of the gross pay is taken for deductions.

▶ **EXAMPLE**

What is the total amount taken from Ariel's gross pay for deductions each month?

STEP 1 Add to find the total percent of deductions.

12% + 8% + 15% = 35%

STEP 2 Find 35% of Ariel's gross pay.

0.35 × $400 = $140

The total amount taken from Ariel's gross pay each week is $140.

Practice and Apply

Use the circle graph to solve each problem. Show your work.

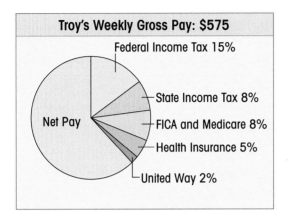

Troy's Weekly Gross Pay: $575

Federal Income Tax 15%

State Income Tax 8%

FICA and Medicare 8%

Health Insurance 5%

United Way 2%

Net Pay

1. How much does Troy pay for health insurance each week?

2. How much does Troy give to United Way each week?

3. What percent of Troy's gross pay is his net pay?

4. **CRITICAL THINKING** Do Troy's deductions added together equal more than $\frac{1}{4}$ of his gross pay? Explain. (Hint: $\frac{1}{4} = \frac{25}{100}$)

Maintaining Skills

Find each percent.
1. 35% of $25   2. 1% of $40   3. 55% of $30   4. 100% of $500

5. 8% of $15.50   6. 20% of $20   7. 95% of $9   8. 75% of $100

4·4 ▸ Focus: Finding Percent, Given the Part and the Whole

You may want to know what percent of your gross pay is taken for deductions. Remember that deductions are taken from your gross pay. You can divide to find a percent if you know the part and the whole.

▸ **EXAMPLE 1**

Stanley's gross pay is $1,000. His deductions are $300. What percent of his gross pay is taken for deductions?

STEP 1 To find the percent, first divide the part by the whole.

$$\frac{\$300}{\$1,000} = \$300 \div \$1,000 = 0.3$$

STEP 2 Change the decimal to a percent. Move the decimal point two spaces to the right. Include zeroes if needed.

$$0.3 = 0.30 = 30\%$$

30% of Stanley's gross pay is taken for deductions.

Sometimes the digits after the decimal point do not end. To change the decimal to a percent, round up or round down.

▸ **EXAMPLE 2**

Erin gives $5 each week to United Way. She earns $240 each week before taxes. What percent of her gross pay does she give to United Way?

STEP 1 To find the percent, first divide the part by the whole. Then round if needed.

$$\frac{\$5}{\$240} = \$5 \div \$240 = 0.02083\ldots$$

$$0.02083\ldots = 0.02 \text{ to the nearest hundredth}$$

STEP 2 Change the decimal to a percent.

$$0.02 = 2\%$$

Here's a Tip!
To round a decimal to the nearest hundredth, look at the third digit to the right of the decimal point. Then round up or round down. 0.666… is about 0.67. 0.333… is about 0.33.

Erin gives about 2% of her gross pay to United Way.

Skills Practice

Divide to find the percent.

1. $\frac{1}{10}$

2. $\frac{25}{1,000}$

3. $\frac{55}{100}$

4. $\frac{1}{5}$

5. $\frac{12}{25}$

6. $\frac{28}{50}$

Write each decimal as a percent.

7. 0.56

8. 0.84

9. 0.7

10. 0.22

11. 0.375

12. 0.804

Write each as a percent.

13. $45 out of $500

14. $81 out of $900

15. $14 out of $350

16. $36 out of $360

17. $15 out of $50

18. $6 out of $200

Everyday Problem Solving

Use the information in the chart. Solve each problem. Round to the nearest percent.

Worker	Occupation	Monthly Gross Pay	Deductions
Kim	Custodian	$1,431	$359
Uri	Movie Attendant	$525	$122

1. Kim works as a school custodian. What percent of her gross pay is taken out for deductions?

2. Uri works at the movie theater. What percent of his gross pay is taken out for deductions?

3. **CRITICAL THINKING** Wallace cooks at Tico's Coffee Shop. His gross pay is $1,033.00 a month. He pays $154.95 in federal tax, $64.05 for Social Security, $15.00 for Medicare, and $123.96 for health insurance each month. What percent of his gross pay is taken for deductions?

What Percent of Gross Pay Is Each Deduction?

Fiona works as a travel agent. Each week she gets a paycheck with an earnings statement attached. Fiona can use the earnings statement to find out what percent of her gross pay is taken for each deduction.

Name: Fiona Wilson   **Social Security Number:** 999-99-9999

Gross Pay	Federal Tax	FICA	Medicare	Health Insurance	United Way	Net Pay
$524.00	$78.60	$34.32	$7.60	$12.20	$15.00	$376.28

▶ **EXAMPLE**

What percent of Fiona's gross pay is deducted for health insurance?

STEP 1 Show health insurance as part of gross pay.

$12.20 ← health insurance
$524.00 ← gross pay

STEP 2 Divide health insurance by gross pay.

$12.20 ÷ $524.00 = 0.02 rounded to the
nearest hundredth

STEP 3 Write the decimal as a percent.

0.02 = 2%

About 2% of Fiona's gross pay is deducted for health insurance.

Practice and Apply

The circle graph shows Fiona's weekly gross pay, her deductions, and her net pay. Use the circle graph to solve each problem. Show your work.

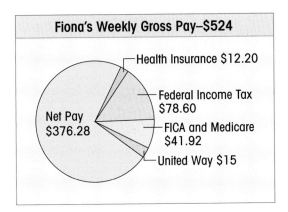

Fiona's Weekly Gross Pay—$524

Health Insurance $12.20
Federal Income Tax $78.60
FICA and Medicare $41.92
United Way $15
Net Pay $376.28

1. Which deduction takes the greatest percent of Fiona's gross pay?

2. Which deduction takes the least percent of Fiona's gross pay?

3. What percent of Fiona's gross pay is taken for federal income tax?

4. What percent of Fiona's gross pay is taken for Social Security and Medicare?

5. What percent of Fiona's gross pay does she give to United Way?

6. **CRITICAL THINKING** Fiona moves to a state that collects state income tax. She finds a new job with the same salary and health benefits. She also continues to give $15 to United Way each week. How will the circle graph change? Explain your thinking.

You have to pay federal income tax on money you earn each year. How much you pay depends on your annual taxable income.

Your income includes your gross pay and other sources of income. **Interest** earned when you keep money in a bank account is also part of your income. Your employer and bank will send you a **W-2 form** each year. The W-2 forms show you the income you have earned.

Your taxable income is your income after deductions and exemptions have been subtracted. Income tax deductions may include donations to charity. Some medical and business expenses may also be deducted. You are allowed one exemption for each dependent.

► **EXAMPLE**

What is Rhonda's taxable income?

Wordwise
A *dependent* is someone related to you or living in your home who *depends* on you for food and housing.

Total Income	
Gross pay	$42,500
Interest from bank account	$110
Exemptions	
2 dependents	$5,800
Deductions	
Business expenses	$4,300
Donation to animal shelter	$200

STEP 1   Add to find Rhonda's total income.
$42,500 + $110 = $42,610

STEP 2   Add to find Rhonda's total exemptions and deductions.
$5,800 + $4,300 + $200 = $10,300

STEP 3   Subtract to find Rhonda's taxable income.
$42,610 − $10,300 = $32,310

Rhonda's taxable income is $32,310.

Practice and Apply

Find each person's taxable income.

	Name	Total Income	Exemptions	Deductions	Taxable Income
1.	Roger	$55,000	$8,700	$8,540 $1,500	?
2.	Tanya	$37,500	$2,900	$2,550 $2,338	?
3.	Gina	$25,400	$5,800	$6,650	?
4.	William	$32,300	$2,900	$3,482 $1,742	?
5.	Paul	$48,000	$2,900	$430 $6,240	?
6.	Marsha	$27,500	$2,900	$5,640 $900	?

7. Leroy makes $41,200 a year. This year, he earned $150 in interest from his savings account. He has $8,700 in exemptions for his dependents. He also has $7,000 in deductions. What is his taxable income?

8. Senja earns $46,200 a year. Her exemptions total $2,900. This year she can deduct $5,000 for business expenses. She also has $2,500 in deductions for medical expenses. What is Senja's taxable income?

9. WRITE ABOUT IT Deductions appear on your earnings statement with your paycheck. You are also allowed deductions when you calculate your annual taxable income. How are these deductions different?

How Do You Read a Tax Table?

You will receive federal income tax forms each year. You will get a tax table with them. You need to read the tax table to find the amount of federal income tax you must pay for the year.

The tax table is organized to show taxable income and filing status. Your status depends on whether you are single, married filing jointly, married filing separately, or the head of a household.

Wordwise

Filing an income tax form means you fill out the form and send it in.

If your taxable income is—		And you are —			
At least	But less than	Single	Married filing jointly	Married filing separately	Head of a household
			Your tax is —		
$28,300	$28,350	4,408	4,249	4,964	4,249
$28,350	$28,400	4,422	4,256	4,978	4,256
$28,400	$28,450	4,436	4,264	4,992	4,264

EXAMPLE

Ellen is single and her taxable income this year is $28,360. How much should she pay this year in federal income tax?

STEP 1   Find the row that has Ellen's taxable income.

Ellen's income is at least $28,350 but less than $28,400.

STEP 2   Find the column for single taxpayers.

Look at the status column labeled *Single*.

STEP 3   Find where the row and the column meet.

Ellen will pay $4,422 for federal income tax.

Practice and Apply

Find the federal income tax each person owes. Use the information in the tax table on page 78.

	Name	Taxable Income	Status	Federal Income Tax Due
1.	Maria	$28,375	Married filing separately	?
2.	Paul	$28,310	Single	?
3.	Lucy	$28,425	Head of household	?
4.	Scott	$28,325	Married filing jointly	?
5.	Debi	$28,435	Single	?

6. **CRITICAL THINKING** John is single. This year his taxable income is $28,325. What percent of his taxable income does he pay for federal income tax? Explain.

Maintaining Skills

Find each percent.
1. 10% of $100
2. 20% of $240
3. 15% of $350

4. 40% of $1,200
5. 25% of $2,500
6. 30% of $900

Write each as a percent.
7. $33 out of $100
8. $8 out of $32
9. $28 out of $70

10. $60 out of $80
11. $99 out of $330
12. $250 out of $500

Is There a Refund or Balance Due?

Money for federal income tax is withheld from each paycheck you receive. What happens if too much money is withheld? What if not enough money is withheld?

If the amount withheld is more than you owe, the federal government will give you a refund. If the amount withheld is less than you owe, you will have to pay the balance to the federal government.

► EXAMPLE

Regina is single. Last year, her taxable income was $21,300. A total of $3,240 in federal income tax was withheld from her gross pay. Will Regina receive a refund or have a balance due?

If your taxable income is—		And you are —			
At least	But less than	Single	Married filing jointly	Married filing separately	Head of a household
		Your tax is —			
$21,250	$21,300	3,191	3,191	3,191	3,191
$21,300	$21,350	3,199	3,199	3,199	3,199
$21,350	$21,400	3,206	3,206	3,206	3,206

STEP 1 Use the tax table to find how much tax Regina owes. $3,199

STEP 2 Compare the amount Regina owes $3,199 < $3,240
and the amount withheld. The amount withheld is greater, so there will be a refund.

STEP 3 Subtract the amount of income tax Regina owes from the amount withheld.

$$\begin{array}{r} \$3,240 \\ -\ \underline{3,199} \\ \$41 \end{array}$$

Regina will receive a refund of $41.

Practice and Apply

Find the refund or balance due for each person. Use the information in the tax table on page 80.

	Name	Status	Taxable Income	Total Federal Tax Withheld	Refund	Balance Due
1.	Richard	Head of household	$21,275	$3,500		
2.	Jeremy	Married filing separately	$21,325	$3,000		
3.	Erin	Single	$21,310	$3,155		
4.	Kirsten	Married filing separately	$21,395	$3,300		
5.	Wesley	Single	$21,375	$3,250		
6.	Mary	Married filing jointly	$21,285	$3,100		
7.	Shanda	Head of household	$21,360	$3,300		
8.	Marvin	Married filing jointly	$21,260	$3,000		
9.	Chou-Li	Single	$21,325	$3,299		

10. IN YOUR WORLD Suppose the federal government owes you a $500 refund. What will you do with your tax refund when you receive it?

What Is Health Insurance?

Many employers offer health insurance. This insurance helps to cover your medical expenses. If you accept this benefit, a deduction may be taken from your gross pay to cover all or some of the cost.

Insurance may cover treatment and medicine you receive from your doctor and your dentist. It may also cover eye care and many hospital expenses. The cost of health insurance depends on the plan.

Did You Know?
HMO stands for Health Maintenance Organization. A traditional plan may be more expensive than an HMO plan.

Health Insurance Plan	Typical Benefits
HMO Plan	• This plan pays the full cost of basic medical needs. • It pays part of many other medical expenses. • There is no limit to the total amount of money you receive each year for medical expenses. • You must use doctors who are listed in the plan.
Traditional Plan	• This plan may limit the total amount of money you receive each year. • Payments for some medical expenses begin only after you have paid a fixed amount called a deductible. • You can receive benefits no matter which doctors you use. • You receive more money for a medical expense if you use a doctor listed in the plan.

The cost of health insurance also depends on how many people your plan covers. The chart below shows some possible costs.

Coverage Choices for Health Insurance	
People Covered	Cost Per Month
Employee	$24.55
Employee and spouse	$28.75
Employee and family	$48.90

Practice and Apply

Use the chart above to solve problems 1–4.

1. Marvin wants medical benefits for himself, his wife, and his three young children. Which insurance coverage should he choose? What is the annual cost? (Hint: Multiply the monthly cost by 12.)

2. Mark is starting a new job and choosing a health plan. What will the health insurance cost Mark each year if he has no spouse or dependents?

3. Agnes is offered health benefits by her employer. Agnes has no children. She is married to an artist who is self-employed. He has no health insurance. Which type of insurance coverage should Agnes choose? What is the annual cost?

4. Marcel is choosing health benefits at a new job. His wife works for another company that offers the same health insurance at a lower cost. What could Marcel and his wife do to get the insurance coverage they need for the least cost?

5. **WRITE ABOUT IT** David has three children and is choosing a health insurance plan. His son Steve has special medical needs. Steve is being treated by a specialist who is not listed with the HMO. Write a short paragraph comparing the HMO and traditional health insurance plans to help David decide.

You can save money in different ways. One way is to use a credit union. A **credit union** offers banking services to company employees at a reduced cost.

You can also have money from your paycheck put directly into a savings account at the credit union. A payroll savings plan allows you to do this.

▶ **EXAMPLE**

Ellie uses a payroll savings plan to deduct $50 each week from her paycheck. The money is put into her savings account at the company credit union. How much money does Ellie save each year?

> Multiply 52 weeks by $50.
>
> 52 × $50 = $2,600

Ellie saves $2,600 each year.

Another way to save money is to buy a **savings bond** with deductions from your paycheck. The government sells Series EE savings bonds at half the face value printed on the bond. You can buy a $100 Series EE savings bond for $50.

Wordwise

To *redeem* a government bond means to sell it back to the government.

When you redeem the bond, you get back more than you paid. The bond increases in value each year until you redeem it. If you keep the bond until it matures, you can sell it for its face value.

Practice and Apply

Solve each problem. Show your work

1. Alric uses a payroll deduction to deposit $35.50 into her savings account each week. How much is deposited into her savings account in one year?

2. Ron wants to save $5,000 in 5 years. Can he do this by having $100 deposited into a payroll savings plan each month? Explain. Use a calculator if you like.

3. How much did Richetta pay for a $1,000 Series EE savings bond?

4. How much is a $1,000 Series EE savings bond worth when it matures?

5. **WRITE ABOUT IT** Karl enjoys buying clothes and CDs. After he pays his bills at the end of each month, he goes shopping. Then he saves whatever money he has left. Write a letter to Karl, explaining why it might be better to have money for savings deducted from his paycheck.

Maintaining Skills

Write each as a percent.

1. $25 out of $100

2. $40 out of $200

3. $4 out of $50

4. $12.15 out of $45

5. $5.50 out of $50

6. $9 out of $45

Compute.

7. $127.00
   −   16.98

8. $145.00
   +   477.27

9. $28.55
   ×   25

10. $145.00
   −   18.58

11. $404
   ×   7.8

12. 16)$15.20

Retirement happens when a person stops working and spends time doing other things. To help you pay your expenses when you retire, you need to plan ahead. Some state and federal agencies put aside some money each pay period for each employee's retirement plan. The money is invested and then given to the employee when he or she retires.

Did You Know?
The part of your gross pay that is put aside for your retirement reduces your taxable income.

Some companies offer a **401(k) plan.** The company deducts money from your gross pay each pay period and puts it into your 401(k) plan. Some companies will also put their own money into your plan as well.

▶ **EXAMPLE**

Eileen's employer adds $0.50 to every $1.00 Eileen puts into her retirement plan. Eileen's monthly earnings statement is shown below. How much money goes into Eileen's retirement plan each year?

Name: Eileen Higgins   **Social Security Number:** 333-33-0003

Gross Pay	Federal Tax	State Tax	FICA	Medicare	401(k)	Net Pay
$2,163	$324.45	$64.89	$134.11	$31.36	$129.78	$1,478.41

STEP 1 Multiply to find the employer's contribution.

$129.78 × $0.50 = $64.89

STEP 2 Add to find the total monthly contribution.

$129.78 + $64.89 = $194.67

STEP 3 Multiply the total monthly contribution by 12.

$194.67 × 12 = $2,336.04

A total of $2,336.04 goes into Eileen's retirement plan each year.

Practice and Apply

Use the information in the monthly earnings statement to solve each problem. Show your work.

Name: Beth Reynolds		Social Security Number: 555-55-5555				
Gross Pay	**Federal Tax**	**State Tax**	**FICA**	**Medicare**	**401(k)**	**Net Pay**
$2,250	$337.50	$67.50	$139.50	$32.62	$112.50	$1,560.38

1. How much is Beth saving for her retirement per year?

2. Beth has a 401(k). Her contribution is the same each month. Beth's employer pays $0.50 for every $1.00 of Beth's contribution. How much is the employer's contribution to Beth's retirement fund each month?

3. Beth's employer pays $0.50 for every $1.00 of Beth's contribution. How much goes into Beth's retirement plan each year?

4. Beth's company changes its retirement plan. Her employer now pays $0.75 for every $1.00 of Beth's contribution. Beth's contribution is the same. Now how much goes into Beth's retirement plan each year?

5. Beth needs to pay an unexpected medical bill. She decreases the amount contributed to her retirement plan each month. She now saves $80 per month instead of $112.50. Beth's employer still pays $0.75 for every $1.00 of Beth's contribution. How much goes into Beth's retirement plan each year?

6. **WRITE ABOUT IT** You have recently received a raise in pay. Why is it a good idea to put some of this money into a retirement plan?

Solve each problem. Show your work.

1. Christine's weekly salary is $358. She pays $53.70 for federal income tax and $17.90 for state income tax. Her FICA and Medicare deductions total $27.39. What is her weekly net pay?

2. Tammy pays 11% of her gross pay for health insurance. Her monthly gross pay is $1,800. How much does health insurance cost Tammy each year? (Hint: There are 12 months in a year.)

3. Rob gives $40 each month to charity. His monthly salary is $1,575. Does Rob give more or less than 2% of his monthly salary to charity?

4. **OPEN ENDED** Marilyn needs to buy two Series EE savings bonds as graduation gifts for her cousins. She has $75 to spend. What could be the face value on each bond? (Hint: The face values of the bonds she can buy are $50, $75, and $100.)

Calculator

You can use a calculator to show $8 out of $12 as a percent.

Press: 8 ÷ 1 2 × 1 0 0 = 66.6666667

So, $8 is about 67% of $12.

Find each percent. Round to the nearest whole percent.

1. $50 out of $125

2. $15 out of $90

3. $7 out of $63

4. $7 out of $70

5. $2 out of $96

6. $88 out of $88

ON-THE-JOB MATH:
Payroll Clerk

Aki is a payroll clerk at a clothing store. She uses a computer program to calculate net pay and print employee paychecks. Aki also pays employees for their business travel expenses.

Aki studied accounting after graduating from high school. She enjoys working with numbers. Aki must check all calculations on employee timesheets and travel reports. She enters the data accurately into the computer. Aki knows that it is important to have the computer records match the data she receives.

Solve each problem.

1. An employee returned from a business trip. This is the record of her travel expenses. How much does Aki enter into the computer for her food expenses?

TRAVEL EXPENSES	
The Eating Palace	$12.45
The Dining Room	$7.90
Lunch Place	$22.00

2. The company pays $0.36 for each mile that employees drive while on company business. An employee drives 25 miles round trip to attend a trade show. How much does Aki record for the mileage expense?

3. The marketing manager asks Aki to deduct $75 from her paycheck each week for deposit into a savings account at the credit union. How much will Aki deposit into the manager's savings account in one year?

You Decide

Aki finds an error in an expense report she receives. The employee recorded a mileage expense based on $0.34 per mile. What do you think Aki should do?

Summary

Federal and state income taxes, FICA, and Medicare are deductions taken from your gross pay.
You can find your net pay by subtracting total deductions from your gross pay.
An earnings statement shows gross pay, deductions, and net pay.
Each deduction is a percent of gross pay. To find the percent, divide the deduction by the gross pay and move the decimal two places to the right.
To find your taxable income, subtract deductions and exemptions from total income.
Use the tax tables to calculate the tax you owe. Sometimes you may have a balance due or you may receive a refund.
Different health insurance plans have different costs and benefits.
You can use payroll deductions to help you save money.
Buying savings bonds is one way to save money for future needs.
A 401(k) plan can help you save money for retirement.

credit union

deduction

exemption

gross pay

net pay

tax

Vocabulary Review

Complete each sentence with a word from the box.

1. An ＿＿＿ is a reduction in your income tax for each person you support, including yourself.

2. Money that is taken from gross pay for union dues is an example of a ＿＿＿.

3. A ＿＿＿ is money paid to the local, state, or federal government.

4. ＿＿＿ is the amount of money received after deductions have been taken.

5. The amount of money earned by an employee is ＿＿＿.

6. A ＿＿＿ is a savings and loan business that offers services to company employees or other groups.

Chapter Quiz

Solve each problem. Show your work.

1. Which payroll deduction might change, depending on which state you live in — federal income tax, state income tax, or Social Security?

2. Why is Harry's net pay for the week less than the amount he earned this week?

3. Julie's gross pay is $1,008.71. Her deductions total $386.00. What is her net pay?

4. Roger's deductions are 44% of his gross pay. What percent of his gross pay is his net pay?

5. Herb and Janice are paid the same salary at a shoe factory. They each work the same number of hours. Herb's net pay is less than the net pay Janice takes home. How could this happen?

6. Carl earns $27,785 a year. He earns $15 in interest on savings. His exemptions and tax deductions total $7,505. What is Carl's taxable income?

7. Magali spent $500 on a Series EE savings bond. She redeemed the bond when it matured. How much did Magali receive for the bond?

Maintaining Skills

Compute.

1. 5% of 135   2. $76.80 + $49   3. 63% of $39

Find the average of each set of numbers.

4. $50.00, $45.25, $33.75, $100.00   5. 5%, 57%, 25%, 33%

Unit 2 Review

Use the circle graph to answer questions 1–5. Write the letter of the correct answer.

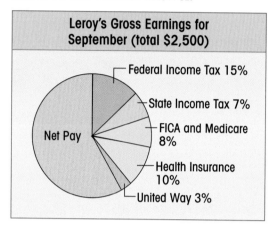

Leroy's Gross Earnings for September (total $2,500)

- Federal Income Tax 15%
- State Income Tax 7%
- FICA and Medicare 8%
- Health Insurance 10%
- United Way 3%
- Net Pay

1. How much state income tax did Leroy pay in September?
   - A. $17.50
   - B. $175
   - C. $191.25
   - D. $375

2. How much was deducted for Social Security in September?
   - A. $191.25
   - B. $252
   - C. $483
   - D. Not enough information is given.

3. How much did Leroy give to the United Way in September?
   - A. $2,500
   - B. $250
   - C. $100
   - D. $75

4. What percent of Leroy's gross earnings did he take home?
   - A. 1%
   - B. 43%
   - C. 57%
   - D. Not enough information is given.

5. Leroy earned a $500 commission in September. What percent of his gross earnings was that?
   - A. 2%
   - B. 20%
   - C. 40%
   - D. 50%

6. Leroy earns $2,000 a month plus a 2.8% commission on sales. His August sales totaled $15,000. What were his gross earnings in August?
   - A. $420
   - B. $2,056
   - C. $2,420
   - D. $17,000

Challenge

In October, Leroy earned more money than he did in September. The amount of money deducted for health insurance does not change. Will the circle graph for his October earnings be the same or different? Explain.

Unit 3 ▶ Banking and Saving

Chapter 5 Choosing a Bank

Chapter 6 Using a Checking Account

A bank can help you manage your money. When you choose a bank, you may want to look for the highest interest rate on checking and savings accounts.

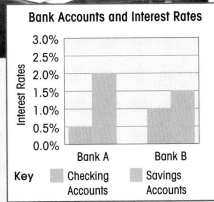

Bank Accounts and Interest Rates

The bar graph shows the interest rates paid by Bank A and Bank B on bank accounts. Use the graph to answer each question.

1. What is the interest rate on the savings accounts offered by Bank A?

2. What is the interest rate on the savings accounts offered by Bank B?

3. Which bank has the better interest rate on checking accounts?

Chapter 5

Choosing a Bank

Compare the services different banks offer. Then choose the bank that best fits your needs. What do you think are some reasons to keep your money in a bank?

Learning Objectives

LIFE SKILLS

- Identify bank services and fees.
- Determine the best bank and checking account.
- Understand and compare savings accounts and CDs.
- Find simple and compound interest.
- Understand how to use an ATM.

MATH SKILLS

- Add, subtract, multiply, and divide with money.
- Find the percent of a number.

Words to Know

check	a piece of paper that orders a bank to pay money from an account
bank	a business that offers financial services
checking account	a bank account against which checks may be written in order to pay expenses
withdraw	to take money out of a bank account
deposit	to put money into a bank account
savings account	a bank account used to save money and earn interest
debit card	a bank card that allows you to use an ATM
automated teller machine (ATM)	a machine that lets you access your bank account at any time
simple interest	interest on the principal only
compound interest	interest on the principal and on the interest already earned
principal	the amount of money in a bank account
certificate of deposit (CD)	a savings account with a minimum balance, a fixed interest rate, and a fixed period of time

Project: Using Banks Now and in the Future

Look in the phone book or on the Internet to find three banks that are near your home, school, or job. Record information for these banks such as the bank's address, phone number, Web site, and services.

How Can a Bank Help You?

You have a job. You work hard and earn an income. Think about the following questions.

- How do you cash your paycheck?
- How do you pay your bills?
- How do you keep your money safe?

You can cash a **check** at a check-cashing service. However, this is expensive.

EXAMPLE

Lola cashed her paycheck at Easy Cash. The fee to cash a check is 5% of the amount of the check. Lola's paycheck was $550. How much was the fee? How much did she receive after she cashed her paycheck at Easy Cash?

Wordwise
A *fee* is an amount of money banks charge you for using a service.

STEP 1	Change the percent to a decimal.	$5\% = 0.05$

STEP 2	Multiply the amount of the check by the decimal.	$\$550 \times 0.05 = \$27.50$

The fee was $27.50 to cash her paycheck.

STEP 3	Subtract the fee from the amount of the check.	$\$550 - \$27.50 = \$522.50$

Lola received $522.50 when she cashed her paycheck.

You can also cash a check at a **bank.** A bank is a business that makes it easy to cash a check, pay bills, and save money. Banks offer many services. Some are listed below.

- You can cash a check at a bank if you have an account there. The money in your account must be greater than or equal to the amount of your check. Banks do not charge a fee for cashing a check.

Here's a Tip!
Some banks give you
access to your account
information on the Internet.

- If you have a **checking account**, you can write and mail a check to pay your bills. The bank will **withdraw** the money from your account.

- You can **deposit** money into a savings account. A **savings account** is a good way to save your money and earn money called interest.

- Banks also offer electronic services. You can use these services even when the bank is closed.

- Some banks have more than one branch or office. Use the branch closest to you if you are in a different town or state. All branches will have access to your account information.

Practice and Apply

Solve each problem. Show your work.

1. Monica cashed her paycheck at Ready Cash. The fee at Ready Cash to cash a check is 10% of the amount of the check. Her paycheck was $538. How much was the fee for Monica to cash her paycheck?

2. Shawn cashed his paycheck at Speedy Cash. The fee at Speedy Cash to cash a check is 9% of the amount of the check. His paycheck was $768. How much did Shawn receive when he cashed his paycheck at Speedy Cash?

3. Holly wants to cash a check for $826 at her bank. She has $535 in her bank account. Will she be able to cash the check at the bank? Explain your answer.

4. Saul wants to cash a check for $715 at his bank. He has $436 in his bank account. How much more money will he need in his bank account to cash the check?

5. **WRITE ABOUT IT** What are some of the reasons you would want to use a bank? Explain.

You have decided to use a bank. Now, you need to find the bank that best meets your needs. Choose the bank that is near your job or home, is open when you need it, and offers the services you need.

Lewis visited each of the three banks near his home. He learned that some banks charge fees for their checking and savings accounts.

Did You Know?
A credit union is a non-profit bank. This means a credit union does not keep its profit.

He learned that he does not have to write a check to pay his monthly car payments. Instead, he can set up electronic withdrawal to pay his car payment automatically on the same day each month.

Lewis discovered that he can use direct deposit to deposit his paycheck automatically instead of driving to the bank every two weeks. Lewis also learned that he can get a **debit card** to use at an **Automated Teller Machine (ATM)** to withdraw money from his bank accounts at any time. Lewis included these services in a chart he made to compare the banks.

Bank Services			
Services	**Dover Bank**	**Newtown Bank**	**Castle Bank**
Monthly service charge for checking accounts	$5	$3	Free
Monthly service charge for savings accounts	$2	$3	Free
Direct deposit	Yes	Yes	Yes
Electronic withdrawal	Yes	No	Yes
Debit cards	Yes	Yes	Yes
Number of branches in state	25	10	5
Open late one night and Saturdays	No	Yes	No

▶ **EXAMPLE**   Lewis wants to open a checking account and a savings account. He wants direct deposit, a debit card, and a choice of branches. Which bank is the best for Lewis?

STEP 1	Add and compare the monthly fees for checking and savings accounts.		Dover Bank	Newtown Bank	Castle Bank
		Checking	$5	$3	Free
		Savings	+ $2	+ $3	Free
		Total	$7	$6	Free

STEP 2	Compare the other services offered.		Dover Bank	Newtown Bank	Castle Bank
		Direct deposit	yes	yes	yes
		Debit card	yes	yes	yes
		Branches	25	10	5

Castle Bank might be the best bank for Lewis. There are no fees. However, there are fewer branches than the other banks.

Practice and Apply

Use the chart on page 98 to solve each problem.

1. Roberta wants her company to electronically deposit her paychecks directly into her checking account. What is the name of this bank service?

2. Juan wants a checking account with direct deposit and a debit card. Which bank is the best for Juan?

3. Su Lee only wants a savings account, a checking account, and electronic withdrawal. She lives close to Newtown Bank and Castle Bank. Which bank is the best for Su Lee?

4. Adam wants a checking account. He wants direct deposit, a debit card, and the bank to be open on Saturdays. Which bank is the best for Adam?

5. **CRITICAL THINKING** Tell why you might want a checking account in Castle Bank.

Julius decides to open a checking account at New City Bank. The bank offers three types of checking accounts. Julius made a chart to compare the accounts.

Checking Accounts at New City Bank			
	Checking Account A	**Checking Account B**	**Checking Account C**
Check fee	None	No fee for the first 8 checks a month then $0.50 a check	$0.40 each check
Debit card fee	No fee at bank's ATM $1.00 each use at any other ATM	No fee at bank's ATM $0.50 each use at any other ATM	At any ATM, no fee for the first 3 times used each month then $0.50 each use
Monthly service charge	$6.00	$4.00	None
Minimum balance required	$100.00	None	None

Julius needs to know about how many checks he will write each month. This will help him to choose the best checking account for him.

▶ **EXAMPLE**

Julius expects to write about 11 checks a month. Which checking account is best for Julius?

Compare the costs for using each checking account.

Checking Account A: Check Fee: Free
Monthly Service Charge: $6.00
Total: $6.00 a month

Here's a Tip!
To find the check fee for Checking Account B, first subtract 8 free checks from the 11 checks a month. Then, multiply $0.50 times 3 checks.

Checking Account B: Check Fee: $1.50 for 11 checks
Monthly Service Charge: $4.00
Total: $5.50 a month

Checking Account C: Check Fee: $4.40 for 11 checks
Monthly Service Charge: Free
Total: $4.40 a month

Checking Account C is best for Julius.

Practice and Apply

Use the chart on page 100 to find the best checking account for each person. Assume each person can keep a $100 minimum balance in the account.

	Name	Number of checks written monthly	Best account
1.	Doug	5	?
2.	Chris	8	?
3.	Carol	11	?
4.	Jen	18	?
5.	Victor	24	?

6. Zack expects to write 20 checks a month. He has Checking Account C. How much will Zack pay for the checking account each month?

7. Adena expects to write 16 checks a month. She has Checking Account B. How much will Adena pay for her checking account each month?

8. Eija expects to write 12 checks a month. How much more a month will it cost him to use Checking Account B than to use Checking Account C? How much more will it cost him in a year?

9. **IN YOUR WORLD** Decide which bills you will pay by check each month when you live on your own. How many checks will you write in one month? Which checking account is best for you? Use a calculator if you like.

How Do You Use an ATM?

What do you do when you need cash and the bank is closed? You can use an Automated Teller Machine, or ATM. The ATM lets you withdraw money from your checking or savings account 24 hours a day. The ATM may only let you withdraw money in multiples of $10 or $20. You can also use an ATM to deposit money into your accounts, or check your account balance.

Here's a Tip!
Do not write your PIN on your debit card! If you lose your wallet, someone could find it and use your debit card.

You need a debit card to use an ATM. You can get a debit card when you open a checking or savings account. When you get this card, you must choose a secret code called a Personal Identification Number, PIN. You must memorize the PIN.

To use the ATM, insert your debit card in the machine's slot. Then, follow the directions on the screen. You will have to enter your PIN.

Your bank may charge a fee when you use an ATM. Do not forget to record each withdrawal and fee.

▶ **EXAMPLE**

Todd used the ATM to withdraw $60. This was the fifth time he used the ATM in May. His bank does not charge a fee for the first two ATM uses each month. After that, the fee is $0.75 for each use. His balance was $487.98 before he used the ATM. What is his new balance?

STEP 1 Decide if there will be an ATM fee. 5 < 2, so yes, $0.75

STEP 2 Add the amount of the cash Todd $60.00
received and the amount of the fee.   +   0.75
                                                  $60.75

STEP 3 Subtract this total from the balance. $487.98
                                                  −   60.75
                                                  $427.23

Todd's new balance is $427.23.

Todd's ATM receipt will show that $60.75 was subtracted from his balance.

Practice and Apply

Solve each problem. Show your work.

1. Phyllis wants to buy shoes for $69. She goes to the ATM to get cash. She must withdraw money in multiples of $20. What is the least amount of money Phyllis should withdraw?

2. Walter used his bank's ATM to withdraw $150. This was the sixth time he used the ATM in June. His bank does not charge for the first five ATM uses each month. After that, the fee is $1.50 each use. What is the total he should subtract from his balance including fees?

3. Kathy used the ATM to withdraw $80. This was the fifth time she used the ATM that month. Her bank does not charge a fee for the first four ATM uses each month. After that, the fee is $0.75 each use. Before she used the ATM, her balance was $845.27. What is the new balance?

4. Ron used the ATM to withdraw $160. This was his second time at the ATM that month. His bank does not charge for the first five ATM uses each month. After that, the fee is $0.75 each use. Before he used the ATM, his balance was $750.20. What is the new balance?

5. **WRITE ABOUT IT** Using an ATM can be convenient, but it can also be expensive. List some things you can do to limit the cost of using an ATM.

Maintaining Skills

Multiply.

1. $1,000 \times 0.015$ 2. $2,000 \times 0.015$ 3. $10,000 \times 0.015$

4. $1,000 \times 0.0125$ 5. $2,000 \times 0.0125$ 6. $10,000 \times 0.0125$

Now Lewis wants to open a savings account to save money. The money the bank pays is called interest. The two types of interest are **simple interest** and **compound interest.**

Simple interest is interest on only the principal. The **principal** is the money you put in the savings account. You can use a formula to compute the interest you will get in your savings account.

Interest = principal × interest rate × time

Time in this formula is in years.

▶ **EXAMPLE 1**

Your wealthy uncle wants to put $16,000 in a savings account with a simple interest rate of 1.5%. How much interest will he earn after 2 years?

STEP 1 Change the percent to a decimal.
1.5% = 0.015

STEP 2 Use the formula to find the interest.
$16,000 × 0.015 × 2 = $480

Your uncle will earn $480 interest after 2 years.

Some savings accounts earn compound interest. With compound interest, you earn interest on the principal and the interest you already earned.

▶ **EXAMPLE 2**

Here's a Tip!
The interest is compounded quarterly, or 4 times a year. Use 0.25 as the time in the formula.

Your uncle decides to put the $16,000 into a savings account with an interest rate of 1.5% that is compounded quarterly. How much interest will he earn in the first quarter? How much interest will he earn in the second quarter?

STEP 1 Use the formula to find the interest for the first quarter.
$16,000 × 0.015 × 0.25 = $60

He will earn $60 interest in the first quarter.

STEP 2 Add the interest to the principal. This is the new principal you will use in the next quarter.
$16,000 + $60 = $16,060

STEP 3 Use the formula again. The interest is compounded. Use $16,060 for the principal for the second quarter.
$16,060 × 0.015 × 0.25 = $60.225

Your uncle will earn about $60.23 interest in the second quarter.

Practice and Apply

Find the simple interest earned.

	Principal	Interest Rate	Period of Time	Simple Interest
1.	$2,500	1.5%	3 years	?
2.	$3,000	3.75%	2 years	?
3.	$4,675	4%	1 year	?
4.	$4,800	3%	5 years	?

 Find the interest earned in the first quarter and the second quarter. Round to the nearest cent. Use a calculator if you like.

	Principal	Interest Rate	When Compounded	First Quarter	Second Quarter
5.	$15,000	2%	Quarterly	?	?
6.	$36,000	3.5%	Quarterly	?	?
7.	$40,000	4.2%	Quarterly	?	?

8. CRITICAL THINKING Adena and Zack each put $1,000 in a savings account at 3.5% for two years. The interest in Adena's account is simple interest. The interest in Zack's account is compounded annually. How much more interest will Zack earn?

5·6 ▶ What Is a Certificate of Deposit?

You can put your savings into a savings account or you can put your savings into a **certificate of deposit (CD)**. A certificate of deposit may offer a higher interest rate than a savings account. You have to keep your money in the account for a fixed length of time. If you take your money out of the account before it matures, you will not receive all the interest your money has earned.

Wordwise
To *mature* means to *reach the maximum value.*

You can use the interest formula to find the amount of interest a certificate of deposit will earn. Then you can find the amount of money that will be in the account when it matures.

▶ **EXAMPLE**

Congratulations! You won first prize of $5,000 in a contest! Of course you want to put this money in an account with the highest interest rate possible. You choose a 6-month CD with an interest rate of 2.6%. The interest is compounded quarterly. How much money will you have in your account when the six months are over?

STEP 1 Multiply the principal by the percent by the period of time. Since the interest is compounded quarterly, 0.25 is the time.

$5,000 × 0.026 × 0.25 = $32.50

STEP 2 Add the interest to the principal.

$5,000 + $32.50 = $5,032.50

STEP 3 Multiply the new balance by the percent by the time.

$5,032.50 × 0.026 × 0.25 = $32.71

STEP 4 Add the interest to the balance.

$5,032.50 + $32.71 = $5,065.21

You will have $5,065.21 in your account.

Practice and Apply

 Find the total value of each CD when it matures. Round to the nearest cent. Use a calculator if you like. (Hint: Semiannually means every 6 months. Use 0.5 for time.)

	Principal	Period of Time	Interest Rate	When Compounded	Value at Maturity
1.	$3,750	3 months	4%	Quarterly	?
2.	$3,000	6 months	3.75%	Quarterly	?
3.	$1,500	6 months	2.5%	Semiannually	?
4.	$4,500	1 year	4.5%	Semiannually	?
5.	$2,500	9 months	4.75%	Quarterly	?

6. IN YOUR WORLD You did well at your job, so you received a bonus of $2,000 last year. You placed the $2,000 in a 1-year CD. The CD has an interest rate of 5%. The interest is compounded quarterly. One year is almost over and you must decide what to do with the money you earned from the interest. How would you use this money? Explain.

Maintaining Skills

Write each percent as a decimal.

1. 5%   **2.** 4%   **3.** 7%   **4.** 9%

5. 7.6%   **6.** 1.3%   **7.** 0.5%   **8.** 0.75%

Find each missing number.

9. 3% of $6,500 is ■.   **10.** 5% of $200 is ■.

11. 2.5% of $1,500 is ■.   **12.** 1.75% of $900 is ■.

Solve each problem. Show your work.

1. Tony cashed a check for $673 at Quick Cash. The fee to cash a check is 11% of the amount of the check. How much did Tony pay to cash his check?

2. Henry wrote 24 checks in April. The bank charges $0.15 a check. The monthly service charge for the checking account is $5. What is the total of the fees Henry paid in April on his checking account?

3. Rex puts $5,000 in a savings account with a simple interest rate of 2.5%. How much interest does Rex earn after 2 years?

4. Beth puts $2,000 in a 6-month CD with an interest rate of 4.5%. The interest is compounded quarterly. How much money will Beth have in her account after six months? Use a calculator if you like.

5. **OPEN ENDED** Jon put $1,000 into each of two different simple interest savings accounts for one year. The total interest earned by both accounts was $5. What could the interest rate have been for each account?

Calculator

Find the simple interest on $500 for six months at the interest rate in each problem below. Round to the nearest cent.

First enter 500 into the calculator's memory. **Press:** [5] [0] [0] [M+] [C] .

Then press [MRC] to retrieve the number for each problem.

1. interest rate of 1.25%   [MRC] [×] [1] [.] [2] [5] [%] [×] [.] [5] [=]

2. interest rate of 2.5%   **3.** interest rate of 5%

DECISION MAKING:
How Should I Manage My Money?

Tim is 24 years old. He has been teaching in the same high school for two years. Tim lives on his own in a one-bedroom apartment in Boston. He does not have a car. He takes public transportation to get to and from work.

Last year Tim saved $1,500. This year, he will be able to save $300 a month except for July and August. Tim does not receive a paycheck in July or August so he uses his savings to pay his expenses for those months.

Use the information above to solve each problem.

1. How much more money can Tim save this year than last year?

2. It is the beginning of the school year. Tim's total expenses for each month are about $1,250. Will he be able to save enough money from September to June to pay his expenses for July and August? If so, how much money will he have left?

3. Tim puts $1,500 from his savings account into a 1-year CD with a simple interest rate of 2.5%. What will the CD be worth in one year?

You Decide

Teachers can often choose how they are paid. One option is to receive a paycheck twice a month for the whole year. The other option is to receive a paycheck twice a month for the school year and not get paid in the summer. If teachers choose not to be paid during the summer, each paycheck during the school year is larger. Either way, they are paid the same total amount. If you were a teacher how would you like to be paid? Why?

Summary

Some of the services banks provide include checking accounts, savings accounts, use of an ATM, and direct deposit.

To decide which checking account is best for you, first decide which bank services you need and the number of checks you would usually write in one month. Then, add up the fees. The total monthly fees will help you decide.

You earn interest when you put money in a savings account or CD.

To find simple interest, multiply the principal by the interest rate by the time.

You need a debit card and Personal Identification Number (PIN) to use an ATM.

ATM
bank
CD
checking account
debit card
principal
withdraw

Vocabulary Review

Complete the sentences with words from the box.

1. A _____ is a savings account with a minimum balance, a fixed interest rate, and a fixed period of time.

2. The amount of money in a bank account is the _____.

3. A bank card that allows you to use an ATM is a _____.

4. An _____ is a machine that lets you access your bank account at any time.

5. A _____ is a business that offers financial services.

6. A _____ is a bank account against which checks may be written in order to pay expenses.

7. To take money out of a bank account is to _____ the money.

Chapter Quiz

Solve each problem. Round to the nearest cent. Show your work.

1. Doug has a checking account. He wrote 10 checks in April. There is no fee for the first 6 checks written in one month. After that, each check is $0.75. He also has a $5 monthly service charge. How much were Doug's bank fees in April?

2. Ling put $1,500 in a savings account. The simple interest rate is 1.5%. How much simple interest will he earn after three years?

3. Juanita put $2,000 in a 6-month CD. The interest rate is 3%. The interest is compounded quarterly. What is the interest for the first quarter? What will be the value of the CD when it matures?

4. Albert used his bank's ATM to withdraw $60. This was the fourth time he used it in June. His bank does not charge a fee for the first two ATM uses in one month. After that, the fee is $0.75 for each use. His balance was $456 before he used the ATM. What is the new balance?

5. Kara put $450 in a savings account with a simple interest rate of 3%. How much interest will she earn after 1 year?

Maintaining Skills

Multiply.

1. $600 × 0.02   2. $700 × 0.04   3. $100 × 0.05   4. $200 × 0.03

Find each missing number.

5. 3% of $6,000 is ▪.

6. 2.5% of $1,600 is ▪.

7. 2.5% of $3,500 is ▪.

8. 4.5% of $3,000 is ▪.

You can put the information in your check register on the computer. The computer will calculate the balance of the money you have in the account. Why is it important to keep track of your balance?

Learning Objectives

LIFE SKILLS

- Describe how to open a checking account.
- Prepare deposit and withdrawal slips.
- Write a check and use a check register.
- Use a debit card.
- Read a bank statement.
- Reconcile a checking account.

MATH SKILLS

- Add, subtract, multiply, and divide with money.

Words to Know

signature card	a card with the name, address, and authorized signature of an account holder
authorized	approved; official
deposit slip	a paper that you fill out when you put money into an account
currency	bills or coins, such as a five-dollar bill; cash
withdrawal slip	a paper that you fill out when you take money out of an account
check register	a booklet used to keep track of the balance when a check is written or a deposit or withdrawal is made from an account
bounce	to refuse to pay a check written for more money than the amount in an account
bank statement	a bank form usually sent each month to the account holder, listing the checks paid, deposits and withdrawals, service charges, and the account balance
reconcile	to compare the information in a check register to a bank statement and correct any mistakes

Project: Interviewing Account Holders

Interview at least ten people you know who have checking accounts.
- Ask each person, *What bank do you use for your checking account?* Use the responses to create a bar graph.
- Ask each person *Why did you choose this bank? (a) convenience, (b) low fees,* or *(c) the services offered?* Use the responses to create a circle graph on a computer.

6·1 ▶ What Does the Bank Need to Know?

You decide you need to have a checking account. You choose a good bank. You decide on the checking account that best fits your needs. You are ready to open your account. What is the first step?

First, the bank will need to know a few facts about you. You must tell the bank if your account is just for you, or for you and another person. By law, the bank must have your Social Security number or your tax identification number. You must show identification, such as a driver's license or a passport, to prove you have given your correct name and address. If you have a job, you must also give the name and address of your employer. You also need to deposit money to open your new account.

Wordwise
Deposit can be used as a noun and a verb. To deposit means to put money into an account. A deposit is the money that you put into that account.

You will be asked to sign a **signature card.** This card shows the bank the **authorized,** or official, way you will sign your checks. The authorized signature is the only signature the bank will accept on your checks. This is done for your protection. If someone steals your checkbook, the signature used will not match yours.

Practice and Apply

Answer each question.

1. What are two things the bank needs from a customer to open a checking account?

2. What number will the bank accept in place of a Social Security number?

3. Why would the bank need your authorized signature?

4. **IN YOUR WORLD** Describe a time when you needed or may need to show identification.

Now that you have a bank account, how do you put money in it? One way is to use an ATM. Another way is to visit the bank and fill out a **deposit slip.** This is Lauren's deposit slip. She is depositing $60 in **currency,** or cash, and a $35 check.

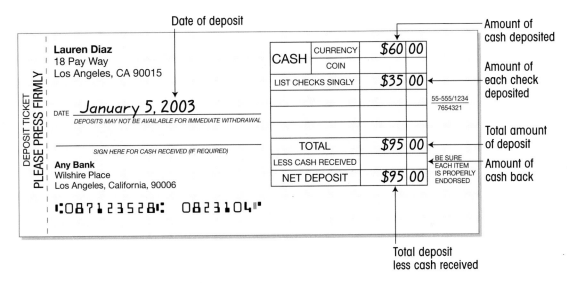

Date of deposit

Amount of cash deposited

Amount of each check deposited

Total amount of deposit

Amount of cash back

Total deposit less cash received

Practice and Apply

Answer each question.

1. How much did Lauren deposit?

2. Did Lauren receive any cash back?

3. What is the date of the deposit?

4. CRITICAL THINKING Lauren wants $25 in cash back. On what line would she write this? Now what is the total of her deposit?

Suppose your friend asks you to go shopping. You have no cash, so you need to withdraw money from the bank. It's easy to withdraw money from your checking account. You can use an ATM machine. You can also visit the bank and fill out a **withdrawal slip.** A withdrawal slip is similar to a deposit slip. Instead of showing how much money you put into your account, a withdrawal slip shows how much money you take out of your account.

Wordwise

A *transaction* occurs any time you deposit or withdraw money. When you withdraw money, the transaction is called a *withdrawal.*

You can receive the money you withdraw in cash or by check. If you are withdrawing a large amount of money, it's better to ask for a check. If you lose the check, no one will be able to cash it without your authorized signature.

Imagine this is your withdrawal slip. You are withdrawing $100 in cash. You fill out the withdrawal slip like the one below.

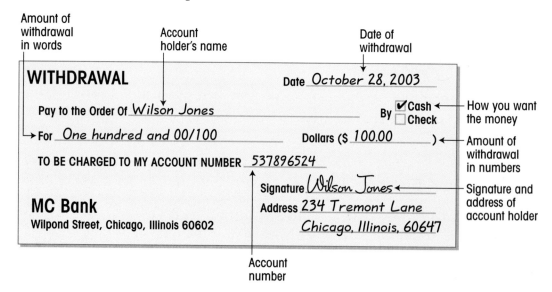

Amount of withdrawal in words

Account holder's name

Date of withdrawal

WITHDRAWAL   Date *October 28, 2003*

Pay to the Order Of *Wilson Jones*   By ☑ Cash ☐ Check

For *One hundred and 00/100*   Dollars ($ *100.00*)

TO BE CHARGED TO MY ACCOUNT NUMBER *537896524*

Signature *Wilson Jones*

MC Bank   Address *234 Tremont Lane*
Wilpond Street, Chicago, Illinois 60602   *Chicago, Illinois, 60647*

How you want the money

Amount of withdrawal in numbers

Signature and address of account holder

Account number

Solve each problem.

```
WITHDRAWAL                          ① Date _____

                                                    ☐ Cash
  ② Pay to the Order Of _____  ③ By  ☐ Check

  ④ For _____  ⑤ Dollars ($ _____ )

       TO BE CHARGED TO MY ACCOUNT NUMBER  537896524

                                    ⑥ Signature _____

  MC Bank                           ⑦ Address _____
  Wilpond Street, Chicago, Illinois 60602
```

1. What are two different methods you can use to withdraw money from your checking account?

2. Imagine you are withdrawing $150 by check from your account today. How would you complete the withdrawal slip above? Copy the slip on your paper and write the missing information for the numbered items.

3. What are the two different ways you can receive the money you withdraw?

4. **WRITE ABOUT IT** Explain whether you prefer to withdraw money from your account using a withdrawal slip or an ATM.

Maintaining Skills

Write each percent as a decimal.

1. 6%   2. 12%   3. 100%   4. 71%

Write each percent as a fraction in simplest form.

5. 25%   6. 4%   7. 30%   8. 98%

Another way to withdraw money is to write a check. A check is a written order to a bank to pay the stated amount of money from an account. When you open a checking account, the bank will give you some blank checks to use. These checks will have your account number on them but not your name and address. You will have to buy checks printed with your personal information. When they arrive in the mail, make sure the information is correct. Here's what to look for:

Did You Know?
Some banks have a Web site you can use to handle many of your transactions online.

• Your name and address are correct.

• The account number on the check matches the account number on your signature card.

• The name and address of the bank are correct.

• The check numbers are in consecutive order.

This is how a check should look when it is filled out properly.

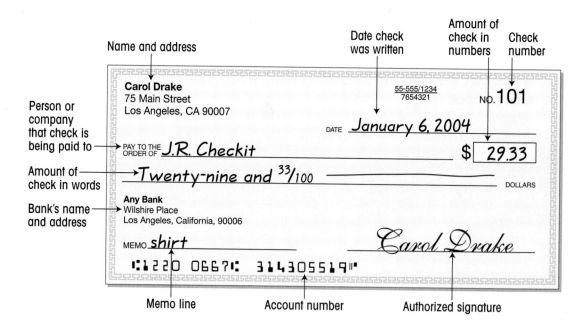

Name and address

Date check was written

Amount of check in numbers

Check number

Carol Drake
75 Main Street
Los Angeles, CA 90007

55-555/1234
7654321

No.**101**

DATE *January 6, 2004*

Person or company that check is being paid to → PAY TO THE ORDER OF *J.R. Checkit* $ 29.33

Amount of check in words → *Twenty-nine and* $^{33}/_{100}$ ————— DOLLARS

Bank's name and address → **Any Bank**
Wilshire Place
Los Angeles, California, 90006

MEMO *shirt*

Carol Drake

⑆1220 0667⑆ 314305519⑈

Memo line

Account number

Authorized signature

Practice and Apply

Today Jason wrote this check to the phone company to pay his telephone bill. Use it to answer each question.

```
╔═══════════════════════════════════════════════════════════════╗
║  Jason C. Tensen                        55-555/1234    NO.109  ║
║  173 Mooney Road                         7654321              ║
║  Parsippany, NJ 07054                                         ║
║  973-555-1234                  ① DATE _____        ║
║                                                               ║
║  PAY TO THE   Northeast Phone Service        ②$[        ]    ║
║  ORDER OF                                                     ║
║       Thirty-two and 76/100 _____ DOLLARS      ║
║                                                               ║
║  U & R Bank                                                   ║
║  Oak Street                                                   ║
║  Parsippany, NJ 07054                                         ║
║                                      Jason C. Tensen          ║
║  MEMO ③ _____                               ║
║   ⑆1220 0667⑆ 314304289⑈                                    ║
╚═══════════════════════════════════════════════════════════════╝
```

1. What information is missing from the check? Write the missing information for the numbered items.

2. What is the amount of the check?

3. What is the check number?

4. What is the account number?

5. **IN YOUR WORLD** List three expenses that you might pay with a check.

Maintaining Skills

Write each of these amounts in words as if you were writing a check.

1. $5.00   2. $10.98   3. $27.86   4. $14.01

5. $100.26   6. $240.89   7. $876.12   8. $1,856.71

6·5 How Do You Keep Track of the Checks You Write?

When you open an account, the bank gives you a **check register** along with your checks. Use it to record each deposit and withdrawal you make and each check you write, as well as any bank fees. This helps you keep track of your balance. To find the balance, add the amount of each deposit and subtract the amount of each withdrawal or bank fee.

Use your check register properly or your checks may **bounce**. This happens when you write a check for more money than you have in your account. Banks charge a fee for this mistake.

This is what a check register looks like.

Wordwise

When you cash or deposit a check, you must write your account number on the back and *endorse* it. To endorse it means to sign your name.

		PLEASE BE SURE TO DEDUCT ANY CHECK CHARGES THAT MAY APPLY TO YOUR ACCOUNT					BALANCE	
NUMBER	DATE	DESCRIPTION OF TRANSACTION	(−) CHECK/DEBIT	✔	(+) DEPOSIT	$	201	00
101	1/6	J.R. Checkit	29 33				−29	33
		for new shirt					171	67
	1/7	Deposit			90 00		+90	00
		yard sale cash					261	67

▶ **EXAMPLE**

Show how to find the final balance on January 7.

STEP 1 Subtract the check amount from the beginning balance.

$201.00 balance
− 29.33 check amount
$171.67 new balance

STEP 2 Add the deposit amount to the new balance.

$171.67 new balance
+ 90.00 deposit amount
$261.67 final balance

The final balance is $261.67.

Practice and Apply

Use Beth Ann's check register to solve each problem.

NUMBER	DATE	DESCRIPTION OF TRANSACTION	(–) CHECK/DEBIT		✔	(+) DEPOSIT		BALANCE $	647	82
206	6/25	Leases-R-Us	279	50						
		car rental								
	6/27	Deposit				550	97			
		paycheck								
207	6/30	P&A Gas & Electric	110	86						
		May utility bill								
	7/2	Withdrawal	50	00						
		for new uniform								

PLEASE BE SURE TO DEDUCT ANY CHECK CHARGES THAT MAY APPLY TO YOUR ACCOUNT

1. What is the balance on each of the four dates?

2. How much money did Beth Ann deposit?

3. What is the number of the check Beth Ann wrote to P&A Gas and Electric?

4. What is the amount of the withdrawal Beth Ann made?

5. On July 5, Beth Ann wrote a check for $217.29. What is her new balance? Use a calculator if you like.

6. **WRITE ABOUT IT** Why do you think it is important to keep track of the checks you write and the deposits you make? What might happen if you kept writing checks until your checkbook was empty?

You learned that you need a debit card to use an ATM. You can also use a debit card instead of writing a check to make a purchase. The amount of your purchase is automatically deducted from your checking account. A debit card is more convenient than writing a check. It is smaller than a checkbook and quicker to use.

Consumer Beware!
A debit card may look like a credit card, but it does not work the same way. A debit card cannot be used to make a purchase for an amount that is greater than the balance in your checking account.

It is a good idea to find out if your bank charges a fee for debit card usage. You should also ask about bank fees for using checks. Compare costs. Sometimes it is cheaper to use your debit card than to write a check.

You should always record debits in your check register. This is how your check register should look when debit card transactions are properly recorded.

NUMBER	DATE	DESCRIPTION OF TRANSACTION	(−) CHECK/DEBIT	✔	(+) DEPOSIT	BALANCE $ 897 61
	8/16	Lowell's Dept. Store	45 99			−45 99
		sweater				851 62
	8/31	ATM (+ $1.25 fee)	121 25			−121 25
		withdrawal				730 37

PLEASE BE SURE TO DEDUCT ANY CHECK CHARGES THAT MAY APPLY TO YOUR ACCOUNT

▷ **EXAMPLE**

Norman's balance was $851.62. He used his debit card to withdraw $120.00 in cash. The ATM fee was $1.25. What is Norman's account balance now?

STEP 1 Add the fee to the withdrawal amount.

$120.00 withdrawal amount
+   1.25 fee
$121.25 total withdrawal

STEP 2 Subtract the total from the account balance.

$851.62 account balance
− 121.25 total withdrawal
$730.37 new balance

Norman's account balance is $730.37 now.

Practice and Apply

This is Indu's check register. Use it to solve each problem.

		PLEASE BE SURE TO DEDUCT ANY CHECK CHARGES THAT MAY APPLY TO YOUR ACCOUNT					BALANCE	
NUMBER	DATE	DESCRIPTION OF TRANSACTION	(−) CHECK/DEBIT		✔	(+) DEPOSIT	$ 1,689	24
142	7/02	Great Skates	149	00			−149	00
		new inline skates					1,540	24
	7/09	Greg's Groceries	18	71			−18	71
		baking supplies					1,521	53
143	7/12	Green County Water	78	65			−78	65
		water bill					1,442	88
	7/20	ATM (+ $1 fee)	81	00			−81	00
		withdrawal					1,361	88

1. On which dates did Indu use her debit card?

2. How do you know when Indu wrote a check rather than used her debit card to make a purchase?

3. What is the amount of the fee that Indu was charged to make a withdrawal from an ATM?

4. On July 20, what is the total amount that was subtracted from Indu's account?

5. **CRITICAL THINKING** Suppose the bank charges $1.50 for each debit card transaction, and $0.35 for each check that is written. Would it always be cheaper for Indu to use her debit card or to write a check? Explain.

At the end of each month, the bank mails a **bank statement** to all account holders. Some banks include your cancelled checks. A bank statement shows which checks have been cashed. It shows withdrawals and use of your debit card. It lists any service charges. It shows the deposits you made.

This is what your bank statement might look like.

	Save-It Bank					
	Fountain Square, Cincinnati, Ohio 45202					
Linda Saylor				Closing Date: 1/28/03		
23 Winters Road						
Cincinnati, Ohio 45202				Beginning Balance: $397.67		

Checking Account Number: 812546982

CHECKS/DEBITS

Check Number	Date Paid	Amount		Check Number	Date Paid	Amount
221	12/29/02	52.98		222	1/11/03	10.23
DEBIT	1/2/03	20.20		224*	1/27/03	11.29

OTHER CHARGES

	Date	Amount
Service Charge	1/1/03	5.00

DEPOSITS

Date	Amount		Date	Amount
1/9/03	174.22		1/20/03	51.76

Ending Balance: $523.95

After you receive your monthly bank statement, you must **reconcile** your account. Compare your register to your bank statement. To reconcile your account, follow these steps.

1. In your check register, check off cancelled checks, deposits, and withdrawals that appear on your bank statement. Make sure all service charges are in the check register. Subtract them from your balance.

2. Find the total amount of outstanding checks and withdrawals or fees not listed on the statement. Then, find the total amount of deposits in your register that are not in your statement.

3. Add the total of all deposits that are not in your statement to your statement's ending balance. Then, subtract the total of the outstanding checks.

4. Your final answer should match your checkbook balance. If it does not, a mistake has been made.

Here's a Tip!
Outstanding checks will not appear on your bank statement; only cancelled checks are listed.

Practice and Apply

Use this check register and the bank statement on page 124 to solve each problem.

NUMBER	DATE	DESCRIPTION OF TRANSACTION	(−) CHECK/DEBIT	✔	(+) DEPOSIT	BALANCE $ 397 67
221	12/29	Feeding U Grocers	52 98			−52 98
						344 69
	1/2	Babbling Books	20 20			−20 20
						324 49
	1/9	Deposit			174 22	+174 22
						498 71
222	1/11	Lakeland Pharmacy	10 23			−10 23
						488 48
	1/20	Deposit			51 76	+51 76
						540 24
223	1/25	Bell Phone Co.	64 12			−64 12
						476 12
224	1/27	Gifts 4 U	11 29			−11 29
						464 83

PLEASE BE SURE TO DEDUCT ANY CHECK CHARGES THAT MAY APPLY TO YOUR ACCOUNT

1. Which entries appear on only one document?

2. List the numbers of any outstanding checks.

3. **CRITICAL THINKING** Reconcile this account using the bank statement. What are the errors, if any?

You have just tried to reconcile your checking account, but your balance doesn't match your bank statement. Now what are you going to do?

Answer these questions to help find the problem.

- Did you check off all debit card uses and cashed checks?
- Did you add and subtract correctly?
- Did you subtract service charges from your balance?
- Did you add all recent deposits to your bank statement? Did you subtract all recent withdrawals from the ending balance on your bank statement?
- Did the bank add and subtract correctly?

Consumer Beware! Some checks come with carbon copies, but these checks are more expensive. You may not want to buy these checks if your bank returns your cancelled checks.

Many bank statements also have a reconciliation form.

Checker Bank • Reconciliation Statement

BALANCE SHOWN ON BANK STATEMENT	$ 747.14	BALANCE SHOWN IN CHECK REGISTER	$ 852.14

PLUS: Deposits in Transit

Date	Amount
3/30	100 00
Total Deposits in Transit	100.00
Subtotal	847.14

PLUS: Corrections

Description	Amount
Total Additions	
Subtotal	0

LESS: Checks Outstanding

Number	Amount
Total Checks Outstanding	0

LESS: Fees and Corrections

Description	Amount
service charge	5 00
Total Deductions	5.00

ADJUSTED BANK BALANCE	$ 847.14	ADJUSTED CHECKBOOK BALANCE	$ 847.14

Use the form to help reconcile your account. If you still cannot find the problem, go to your bank for help. Take your statement and check register with you. The bank staff should be able to help you.

Practice and Apply

Use the bank statement and check register below to answer each question.

Checker Bank
Salem Way, Dover, NJ 07801

Todd Payman
23 Hazelwood Road, Dover, NJ 07801

Closing Date: 3/29/03
Beginning Balance: $702.43

Checking Account Number: 469853478

CHECKS

Check Number	Date Paid	Amount	Check Number	Date Paid	Amount
488	3/10/03	23.39	489	3/28/03	15.90

OTHER CHARGES

	Date	Amount
Service Charge	3/29/03	5.00

DEPOSITS

Date	Amount	Date	Amount
3/17/03	89.00		

Ending Balance: $747.14

PLEASE BE SURE TO DEDUCT ANY CHECK CHARGES THAT MAY APPLY TO YOUR ACCOUNT

NUMBER	DATE	DESCRIPTION OF TRANSACTION	(−) CHECK/DEBIT	✔	(+) DEPOSIT	$ BALANCE	
488	3/8	Video World	23 39	✔		702	43
						−23	39
						679	04
	3/17	Deposit		✔	89 00	+89	00
						768	04
489	3/26	Toy Depot	15 90	✔		−15	90
						752	14
	3/30	Deposit			100 00	+100	00
						852	14

1. What charge did Todd forget to include in his check register?

2. What is the closing date of the bank statement? What is the date of the last transaction in the checkbook?

3. **WRITE ABOUT IT** Explain why the ending balance on the bank statement differs from the last balance in the checkbook.

Solve each problem. Show your work.

1. Charlie is a landscaper. He needs a new lawn mower that costs $545. He had $150 in his checking account. Then he deposited checks for $120, $46, $97, $112, and $88. If he buys the lawn mower, how much will he have left in his account?

2. Carrie thought she had $90 in her checking account. But, she forgot to record checks she wrote for $45 and $20. What is Carrie's balance?

3. Hamad had $518.76 in his checking account. His bank bounced his check for $525.00. How much more did Hamad need in his account before writing the check?

4. **OPEN ENDED** Aldo made four transactions in May. The beginning balance on his bank statement is $672.08. The ending balance is $547.63. A check for $150.00 was cashed. A deposit for $87.25 was made. What could the other two transactions have been?

Calculator

A calculator is helpful when reconciling your account.

Enter the beginning balance.		$402.31	beginning balance
Add deposits.	+	39.00	deposit
Subtract checks and charges.	−	108.92	check
Press = for the new balance.		*332.39*	new balance

Use a calculator to solve this problem.

Your beginning balance is $87.01. You make a deposit of $53.27. You write checks for $32.48 and $11.27. What is your new balance?

ON-THE-JOB MATH:
Bank Teller

Elena is a bank teller. She has a high school diploma and has been working as a teller for three years. She likes helping people and working with numbers.

Elena makes sure that customers write the correct amounts on deposit and withdrawal slips. She always counts the money she gives to customers at least twice before handing it to them. At the end of the day, Elena makes sure that the money in her cash drawer is correct, based on her computer record.

One day, Elena started with $1,000.00 in her drawer. That day, she accepted $7,842.67 in cash deposits. She paid out $3,080.00 in cash. According to her computer, at the end of the day she should have had $5,762.67 in her drawer. Elena counted only $5,722.67 in cash. The table lists the cash that she had in her drawer.

Use the chart to answer each question.

1. How much money in coins is in Elena's drawer?

2. How much money in bills is actually in her drawer?

3. What is the difference between the total in her computer and the actual amount in the drawer?

4. What error did Elena make?

Currency	Quantity
$100 bills	28
$50 bills	20
$20 bills	50
$10 bills	60
$5 bills	40
$1 bills	120
quarters	83
dimes	128
nickels	149
pennies	167

You Decide

Suppose you are a bank teller. How would you make sure that you counted the money in your drawer correctly at the end of each day? How many times would you count it? Would you use a calculator?

Summary

To open a checking account, you need to give the bank certain information, fill out a signature card, and make a deposit into your account.

You fill out a deposit slip each time you make a deposit to your account. All cash and check amounts must be listed. All checks must be endorsed with the account number.

You fill out a withdrawal slip each time you take money out of your account at the bank. You must choose how you would like to receive the money you withdraw.

You must be sure that the information on your printed checks is correct. When you write a check, include the check amount in words and numbers, the date, the person or company you are writing the check to, and your signature.

It is important to enter each transaction into your check register and find the new balance.

A debit card can be used to pay for a purchase.

Each month, you will receive a bank statement to help reconcile your account. If your checking account doesn't reconcile, there are several steps you can take to find the error.

check register

currency

deposit slip

reconcile

signature card

withdrawal slip

Vocabulary Review

Complete the sentences with words from the box.

1. A paper that you fill out when you take money out of an account is a _____.

2. When you compare the information in a check register to a bank statement and correct any mistakes, you _____ your account.

3. Money in bills or coins is called _____.

4. A _____ is a booklet used to keep track of the balance when a check is written or a deposit or withdrawal is made from an account.

5. A paper that you fill out when you put money into an account is a _____.

6. A _____ is a card with the name, address, and authorized signature of an account holder.

Chapter Quiz

Solve each problem.

1. Ezekiel is opening a checking account. He gave the bank his information and filled out a signature card. What else does he need to do?

2. Lola is depositing a check for $168. She wants to receive $68 back in cash. Copy and complete the last two lines of the deposit slip to show the transaction.

3. Ray has $432.76 in his checking account. He makes a withdrawal of $60.00. What is his account balance now?

CASH		
LIST CHECKS SINGLY	168	00
TOTAL FROM OTHER SIDE		
TOTAL	168	00
LESS CASH RECEIVED		
NET DEPOSIT		

4. What is the balance on each of the three dates shown on the check register below?

		PLEASE BE SURE TO DEDUCT ANY CHECK CHARGES THAT MAY APPLY TO YOUR ACCOUNT							
NUMBER	DATE	DESCRIPTION OF TRANSACTION	(−) CHECK/DEBIT		4	(+) DEPOSIT		$	BALANCE 592 11
463	7/25	CD Universe	14	78					
	7/27	Deposit				112	64		
464	7/30	Save the Beaches	25	00					

5. Shirley is reconciling her checking account. She has compared the ending balance in her bank statement to the balance in her check register. They both say $430.29. Is Shirley finished reconciling her account? Explain.

Maintaining Skills

Find each percent.

1. 13% of 100
2. 25% of 268
3. 150% of 100
4. 4% of 325

5. 0.5% of 100
6. 12% of 12
7. 1% of 1
8. 6.5% of 524

Unit 3 Review

Use the bar graph to answer questions 1–2. Write the letter of the correct answer.

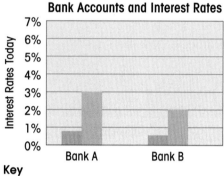

Bank Accounts and Interest Rates

Key
- Checking Accounts
- Savings Accounts

1. Sarah put $2,000 into a savings account in Bank B. How much simple interest will she earn after three months?

    A. $2.00
    B. $10.00
    C. $12.50
    D. none of the above

2. Tomas put $1,000 into a savings account in Bank A. What will his balance be in 3 months?

    A. $7.50
    B. $30.00
    C. $1,000.00
    D. $1,007.50

3. What is the yearly simple interest on $1,800 if the rate is 4.5%?

    A. $72.00
    B. $81.00
    C. $810.00
    D. $1,881.00

4. Al's bank charges $0.12 for each check he writes. Al wrote 17 checks in May. How much will this cost?

    A. $0.34
    B. $0.84
    C. $2.04
    D. $4.02

5. Maryanne had $362 in her checking account. She wrote a check for $40 and deposited $25. What is her new balance?

    A. $297
    B. $347
    C. $374
    D. $377

Challenge

Singh invested his money in an account with a simple interest rate of 2.5%. Ida invested her money in an account with a simple interest rate of 2.0%. Could Ida earn more interest than Singh in one year? Explain.

Chapter 7 **Credit Card Math**

Chapter 8 **Loans and Interest**

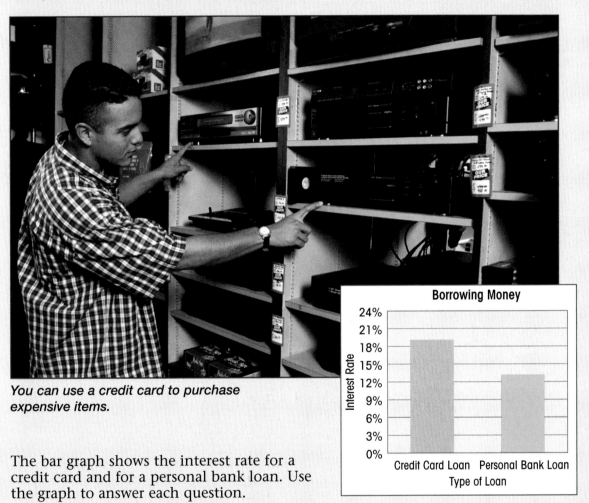

You can use a credit card to purchase expensive items.

The bar graph shows the interest rate for a credit card and for a personal bank loan. Use the graph to answer each question.

1. What is the interest rate on a credit card?

2. What is the interest rate on a personal loan?

3. Does the credit card or the personal loan have a higher interest rate?

Careless use of credit cards can ruin your budget. What should you think about before you use a credit card?

Learning Objectives

LIFE SKILLS

- Compare the good and bad things about credit cards.
- Understand how to complete a credit card application.
- Identify ways to protect a credit card account.
- Find the amount of a finance charge.
- Read and understand a monthly statement.
- Find the new balance on a credit card account.
- Decide if a credit card should be used to buy items.
- Find how a cash advance affects your credit card account.

MATH SKILLS

- Find the percent of a number.
- Find the percent, given the whole and the part.

Words to Know

credit card	a card that allows you to buy items now and pay for them later
monthly statement	a form like a bank statement sent by your credit card company to bill you for what you owe every month and to update you on the status of your account
minimum payment	the least amount of your balance that you must pay to the credit card company each month
co-applicant	a person who applies with you for a credit card account
balance	the total amount of money you owe on your account
finance charge	a fee you pay each month for borrowing money; it is based on a certain percent of the balance not paid
annual percentage rate (APR)	yearly interest rate used to find the interest due each month on the balance not paid
credit limit	the most money you can borrow on a credit card
available credit	the amount you can still charge on your card; the difference between the credit limit and the balance on your account
credit	the amount of a return or payment on a monthly statement
cash advance	the cash borrowed from a credit card account

Project: Comparing Credit Cards

Compare the advantages of a credit card from a national credit card company with the advantages of a credit card from a large store in your area. You can use the Internet to find the latest information on both credit cards. Consider interest rates, convenience, and special offers.

Write a summary report about the advantages you found for both credit cards and which you prefer. Include a bar graph to compare the interest rates.

How can a **credit card** be useful? You can use a credit card almost anywhere you go. You can pay with a credit card instead of using cash or a debit card. A credit card lets you buy something now and pay later.

Did You Know?
Most credit card companies and banks allow you to have Internet access to your credit card account information on a secure Web site.

You can open a credit card account with a bank, credit card company, or store. You are given a set amount of credit that you can use to buy things. When you use a credit card, you are really borrowing money.

You will get a **monthly statement** for your credit card. You can pay part or all of the balance you owe each month. You must pay at least the **minimum payment** by the due date. If you don't pay the entire bill, you will pay interest on the balance. If your payment is late, you will be charged a late fee.

The charts below list some good and bad things about credit cards.

Good Things About Credit Cards
• Your monthly statement is a record of the items you buy.
• You can buy things by using the telephone or the Internet.
• You can pay for things in an emergency.

Bad Things About Credit Cards
• You have to pay interest if you don't pay the entire balance.
• The money you pay in interest increases the cost of what you buy.
• You may buy more than you can afford.

Practice and Apply

Answer each question.

1. If John pays the entire balance on every monthly statement, will he owe any interest? Explain.

2. Does using his credit card lower the amount of credit John has left to use in that account? Explain.

3. **IN YOUR WORLD** Suppose you have a credit card. How can you avoid using it to buy more than you can afford?

How Do You Open a Credit Card Account?

How do you open a credit card account? You can fill out an application and mail it or bring it to the store or bank. You can also apply by phone or on the Internet. Most credit card applications ask for the same information. However, one application may look different from another. Look at the application on page 138 to see what information you need to give.

You may want to apply for a joint credit card account with another person. You must give information about this **co-applicant** on the application.

Practice and Apply

Use the credit card application on page 138 to answer each question.

1. John's birthday is September 27, 1980. How will he write the date of his birthday on the application?

2. John has lived in the same place for 27 months. Would he put his previous address on this application? Explain.

3. John works as a bank teller. Where should he put this information?

4. John has had his present job for 28 months. What would he write above "How Long?" in the employer section?

5. John's father is his co-applicant. What type of account will John check on the application: individual or joint?

6. If John has a checking account, does he need to write this information on the application? If so, where?

7. **WRITE ABOUT IT** Find a business in your town that offers credit cards. Write a formal letter requesting a credit card application.

Credit Card Application

Please print and provide all requested information below.

FIRST NAME	INITIAL	LAST NAME

/ /	– –		
DATE OF BIRTH	SOCIAL SECURITY NUMBER	DRIVER'S LICENSE NUMBER	STATE

MOTHER'S MAIDEN NAME

☐ RENT ☐ OWN

HOW LONG? _____ YRS. _____ MOS.

HOME PHONE ( ) –

CURRENT HOME ADDRESS	APT#	CITY	STATE	ZIP

PREVIOUS HOME ADDRESS (If less than 2 years at current address)	APT#	CITY	STATE	ZIP

		YRS. _____ MOS.
EMPLOYER NAME	POSITION	HOW LONG?

BUSINESS ADDRESS

( ) –

BUSINESS PHONE

FORMER EMPLOYER (If less than 2 years at current employer)

TYPE OF ACCOUNT REQUESTED (CHECK ONE): ☐ INDIVIDUAL ☐ JOINT

CO-APPLICANT INFORMATION (if Joint Account requested)

FIRST NAME	INITIAL	LAST NAME

	– –
RELATIONSHIP TO APPLICANT (if any)	SOCIAL SECURITY NUMBER

		YRS. _____ MOS.
EMPLOYER NAME	POSITION	HOW LONG?

( ) –	( ) –
EMPLOYER ADDRESS	BUSINESS PHONE HOME PHONE

BANK—LIST BRANCH AND ADDRESS

☐ CHECKING
☐ SAVINGS

OTHER CREDIT REFERENCES? 　　　　　　　　　　ACCOUNT NO.

7·3 How Do You Protect Your Credit Card?

Now that you have a credit card, you need to protect it just like your cash or your keys.

If your credit card is lost or stolen, call the credit card company's customer service department immediately. The phone number is on your monthly statement. Federal law says that you pay at most $50 if someone steals your credit card and uses it to buy something.

Some more tips are in the chart below.

Tips for Protecting Credit Cards
1. Sign the back of the credit card as soon as you get it in the mail.
2. Do not give your credit card account number to any strangers.
3. Only shop on secure Web sites that will protect your credit card information.
4. Do not let your friends or family use your credit card. You are responsible for paying this bill, even if they do not pay you back.
5. Compare your credit card receipts to your monthly statements. If you find an item listed that you did not buy, write or call the credit card company immediately. Be sure to tell them that this is a billing error.

Practice and Apply

Answer each question below.

1. Jen lost her wallet. Her credit card was in her wallet. What should she do first?

2. The credit card Sal applied for just arrived in the mail. What should he do first?

3. Someone steals your credit card and uses it to buy things. How much do you have to pay?

4. **WRITE ABOUT IT** Sometimes people try to get your credit card information to use illegally. Learn about consumer protection at the library or at www.bbb.org. Click on "Consumer Information." Then, click on "Scams." Describe what you learn.

What About Buying on Credit?

You opened your credit card account. Now, you can use your credit card to buy things. Your monthly statement will show all the items you bought last month.

Each month you must send the credit card company a payment. If you do not pay the entire **balance**, you can send a payment for less. But you must send at least the minimum payment. If you do not pay the entire balance, you will owe a **finance charge** on the unpaid balance.

An **annual percentage rate (APR)** is used to compute the new finance charge each month.

▶ **EXAMPLE**

Sean used a credit card to buy new clothes for his job interview. The new clothes cost $220. The next monthly statement on Sean's credit card showed an unpaid balance of $220. Sean paid the $20 minimum payment. The annual percentage rate for the card is 18%. What will the finance charge be for one month on the unpaid balance?

Did You Know?
Your credit card company must receive your payment by the due date to avoid a late charge. To make sure this happens, mail your payment at least 7 days before the due date.

STEP 1   Find the monthly percentage rate.
Change the APR to a decimal number.   18% = 0.18
Divide the APR by 12.   0.18 ÷ 12 = 0.015

STEP 2   Find the unpaid balance. $220 − $20 = $200
Subtract the payment from the total bill.

STEP 3   Multiply the unpaid balance $200 × 0.015 = $3
by the monthly percentage rate.

The finance charge for one month on the unpaid balance is $3.

Practice and Apply

 Find each monthly percentage rate. Then calculate the finance charge for one month on the unpaid balance. Use a calculator if you like.

	Name	Unpaid Balance	Annual Percentage Rate	Monthly Percentage Rate	Finance Charge
1.	Cherice	$100	12%	?	?
2.	Max	$350	12%	?	?
3.	Dave	$450	18%	?	?
4.	Ralph	$500	18%	?	?
5.	Janice	$900	24%	?	?

Solve each problem. Use a calculator if you like.

6. Connie's credit card has an unpaid balance of $750. The annual percentage rate is 18%. What will be the finance charge for one month on this amount?

7. Ralph had a $400 unpaid balance after last month's payment on his account. His annual percentage rate is 24%. The minimum payment due is $20. How much of that is interest for one month on the unpaid balance?

8. CRITICAL THINKING Crystal has an unpaid balance of $620 on her credit card. She decides to pay more than the minimum payment due on the account. How does that affect the finance charge on next month's statement? Explain.

How Do You Read a Monthly Statement?

Here's a Tip!
Mail the top part of your monthly statement with your payment to the credit card company. Keep all monthly statements and credit card receipts as records.

Look at the monthly statement on page 143. A monthly statement shows all the activity on an account for the past month. A charge is any activity, such as a purchase, that increases your unpaid balance. A credit is any activity, such as a payment, that reduces your unpaid balance.

The monthly statement also shows your **credit limit** and how much **available credit** you have left.

Practice and Apply

Use Sean's monthly statement on page 143 to answer each question.

1. What is Sean's credit card account number?

2. When is the minimum payment due?

3. How much is the minimum payment due?

4. If Sean uses a check to make his payment, to whom does he write the check?

5. When was Sean's last payment received?

6. What is the annual percentage rate on purchases?

7. What is the total amount of the charges?

8. If Sean has a question about his monthly statement, what is the phone number he calls?

9. Can Sean buy a computer that costs $1,500 with this credit card account? Explain.

10. **CRITICAL THINKING** If Sean had sent in a payment of $150 instead of $100 on April 23, what would change on the account summary?

PRIME CREDIT COMPANY

WORK ( )
HOME ( )

YOUR ACCOUNT NUMBER: 123-45-670

PAYMENT DUE DATE	YOUR NEW BALANCE	MINIMUM PAYMENT DUE	ENTER AMOUNT OF PAYMENT ENCLOSED
06/15/03	$145.15	$10.00	$, .

Return this part with your payment.

Please make check payable to Prime Credit Company.
Be sure to write your account number on the check.

‖‖ɪɪɪ‖ɪɪɪɪ‖‖ɪɪɪ‖ɪ‖ɪɪ‖ɪɪɪ‖ɪ‖‖ɪɪɪ‖ɪ‖‖

Prime Credit Company
P.O. Box 061788
Camden, NJ 08104-4321

Sean Davis
12 East Broadway Street
Galloway, IL 55555-1234

ACCOUNT NUMBER: 123-45-670

STATEMENT CLOSING DATE	DAYS IN BILLING CYCLE
05/20/03	30

CREDIT LIMIT	AVAILABLE CREDIT	CASH ADVANCE LIMIT	AVAILABLE CREDIT FOR CASH ADVANCE
$1,000.00	$854.85	$100.00	$100.00

CHARGES AND CREDITS

Activity Date	Post Date	Reference Number	Activity	Amount
04/23	04/23	1C2D	Payment	– 100.00
04/25	04/26	2E4J	Computer Software	+ 40.00
	05/20		Finance Charge	+ 2.15

For Finance Charge	Annual Percentage Rate
Purchases	18.0%
Cash Advances	19.0%

ACCOUNT SUMMARY:

Previous Balance	Total
	$203.00
(–) Payments, Credits	– 100.00
(+) Purchases, Cash, Debits	+ 40.00
(+) FINANCE CHARGES	+ 2.15
(=) New Balance	$145.15
Minimum Payment Due	$10.00

For answers to any questions, please call us at 1-888-555-1234.

What Happens When You Return an Item?

You buy an item and later decide you do not want to keep it. Lucky for you that you kept the receipt! You will need this record of your purchase to return an item.

When you return an item, bring the item and the receipt to the store. If you used a credit card to pay for the item, the store puts a **credit** on your account. This means that you no longer owe the money for the item you returned. Keep a copy of the new receipt with the credit to prove you returned the item to the store. Then, check your next monthly statement to be sure that the credit is recorded correctly.

▶ **EXAMPLE**

Wordwise
A *previous balance* is the old balance on your monthly statement before any new activity is recorded.

Last month, Susan used her credit card to buy a $600 computer and a $150 printer. Later that month, Susan returned the printer. When Susan got her next monthly statement, she saw that there was a previous unpaid balance of $80. The statement also showed a credit for the printer and a payment of $300. The finance charge for the month was $1.20. What should the new balance be?

STEP 1   Find the total for new charges. Add the charges for the computer, the printer, and the finance charge on the previous balance.

$600 + $150 + $1.20 = $751.20

STEP 2   Add the total charges to the previous balance.

$751.20 + $80 = $831.20

STEP 3   Find the total credits. Add the payment to the return credit.

$300 + $150 = $450

STEP 4   Find the new balance. Subtract the total credits.

$831.20 − $450 = $381.20

The new balance should be $381.20.

Practice and Apply

 Find each new balance. Use a calculator if you like.

	Previous Balance	New Purchases	Finance Charge	Credits	Payment	New Balance
1.	$100	$150	$3.00	None	$50	?
2.	$90	$225	$3.30	$50	$100	?
3.	$120	$370	$3.60	$100	$30	?
4.	$50	$400	$3.50	$80	$20	?
5.	$200	$400	$5.85	$60	$150	?
6.	$500	$500	$6.13	None	$10	?

Solve each problem. Show your work.

7. Manisha used her credit card to buy school books for $400. Later, she returned one book for $50. Manisha had a previous balance of $70. When Manisha got her new monthly statement, it showed a payment of $250 and a finance charge of $1.05. What will be the new balance on the statement?

8. Terry used his credit card to buy two suitcases. Together, they cost $350. He later returned one suitcase for $140. Terry had a previous balance of $230. Terry's next monthly statement showed a $200 payment. It also showed a finance charge of $3. What will be the new balance on the statement?

9. **CRITICAL THINKING** Dillan used his credit card for the first time. He spent $500 on clothes and shoes for work. The minimum payment on his account is always 2.5% of the unpaid balance. How much is the first minimum payment? (Hint: 2.5% = 0.025)

7·7 Should You Continue to Buy on Credit?

Suppose you have more than one credit card. Be careful! It is easy to buy more than you can afford when using credit cards.

Did You Know?
It is good to have enough money in savings to pay for at least two months of expenses.

How do you know when to stop buying on credit? You should be saving a part of every month's pay. You should also keep the credit card balances low enough so that you could pay the entire balance without too much difficulty. The total of your monthly credit card payments should be no more than 20% of your monthly take-home pay.

▶ **EXAMPLE**

Your take-home pay is $2,000 every month. Your credit card payments are about $280 every month. About what percent of your take-home pay is used to make your credit card payments? Should you continue to buy on credit?

STEP 1 Divide the total credit card payments by the take-home pay.

$280 ÷ $2,000 = 0.14

STEP 2 Change the decimal to a percent.

0.14 = 14%

Your credit card payments are about 14% of your take-home pay. You can continue to buy on credit.

Sometimes unexpected things happen. You should pay attention to how you are using your credit cards. Plan your finances wisely. Do not let your credit card debt become more than you can afford.

Practice and Apply

Find each percent of take-home pay. Then decide if buying
on credit is a good idea for each person. Write *yes* or *no*. (Hint:
Payments should be no more than 20% of take-home pay.)

	Name	Total Monthly Credit Card Payments	Monthly Take-home Pay	Percent of Take-home Pay	Is this person buying too much on credit?
1.	Mari	$357	$1,700	?	?
2.	Jude	$250	$1,000	?	?
3.	Karib	$190	$1,000	?	?
4.	Aaron	$216	$1,200	?	?

Solve each problem. Use a calculator if you like.

5. Calvin's monthly take-home pay is $1,050. What is the
most he should be paying each month on credit cards?

6. WRITE ABOUT IT Lynn's take-home pay is $882 every
month. Her credit card payments are about $150 each
month. She plans to use her credit card to pay for a
computer course. This will increase her monthly credit
card payments to $285 each month. Should Lynn use her
credit card to pay for the computer course? Explain.

Maintaining Skills

Compute.

1.	$395.85	**2.**	$472.99	**3.**	$98.00	**4.**	$602.15
	+ 175.32		+ 588.51		− 42.36		− 234.78

5. $32.75 × 17 =

6. $50.25 × 20 =

7. $45.75 ÷ 3 =

8. $33.12 ÷ 24 =

7·8 ▷ What Is a Cash Advance?

You know how to use your credit card to buy things. You can also borrow money with your credit card. You can use your credit card at any ATM to get a **cash advance**. You can get as much money as your cash advance limit allows.

The annual percentage rate for cash advances is often higher than the annual percentage rate for purchases.

▶ **EXAMPLE**

Mari gets a $400 cash advance on her credit card. The annual percentage rate for cash advances is 24%. What is the finance charge for one month on the cash advance?

STEP 1 Find the monthly percentage rate on a cash advance. Change 24% to a decimal. Then divide by 12.

24% = 0.24

0.24 ÷ 12 = 0.02 = 2%

STEP 2 Find the finance charge on the cash advance. Multiply the cash advance by the monthly percentage rate.

$400 × 0.02 = $8

The finance charge for one month on the cash advance is $8.

Sometimes the balance on a credit card includes charges for purchases as well as cash advances. A separate finance charge is calculated monthly for each type of charge. The total finance charges are added to the balance.

Practice and Apply

Solve each problem. Show your work.

1. The annual percentage rate on a cash advance is 18%. What is the finance charge for one month on a $250 cash advance?

2. Yang gets a $550 cash advance on his credit card. The annual percentage rate for cash advances is 24%. What is the finance charge for one month?

3. The annual percentage rate on a cash advance is 24%. The APR on a credit card purchase is 12%. How much more interest do you pay in one month on a cash advance of $500 compared to a $500 purchase?

Find the total finance charge for the month. (Hint: First find the finance charge for the cash advance. Then find the finance charge for the purchase. Then add.)

	Cash Advance	Cash Advance APR	Amount of Purchase	Purchase APR	Total Finance Charge
4.	$800	24%	$300	18%	?
5.	$500	18%	$150	12%	?
6.	$500	18%	$60	12%	?
7.	$500	24%	none	18%	?
8.	$900	15%	$220	12%	?
9.	$900	18%	$400	15%	?

10. **CRITICAL THINKING** Sharon sees a coat on sale for 20% off the original price. She wants to use a cash advance to pay for it. The cash advance has an annual percentage rate of 24%. Is this a good way to buy the coat? Explain.

Solve each problem. Use a calculator if you like.

1. Travis must pay a minimum of 2% of the unpaid balance on his credit card. How much is the minimum payment on a balance of $432?

2. Susan has a balance of $800 on her credit card. The APR is 12%. By how much does a $20 payment reduce the balance? (Hint: Find the interest for one month.)

3. **OPEN ENDED** Jan's monthly take-home pay is $1,050. She has two credit cards. How much can she pay for the two cards every month? (Hint: Total payments should be no more than 20% of her monthly take-home pay.)

Calculator

The APR on your credit card is 24%. Your beginning balance is $300. You make a $20 payment. You can use the memory keys to find the finance charge on the unpaid balance.

First find the monthly interest rate. Store this number in memory.

Press: `.` `2` `4` `÷` `1` `2` `M+` | M   0.02 |

Then press the keys needed to find the unpaid balance. Multiply this by the number you stored in memory. Press MRC.

Press: `3` `0` `0` `−` `2` `0` `×` `MRC` `=` | 5.6 |

Find each finance charge on the unpaid balance.

	APR	Beginning Balance	Payment	Finance Charge on Unpaid Balance
1.	18%	$450	$30	?
2.	20%	$315	$20	?
3.	15%	$980	$120	?

DECISION MAKING:
Which Credit Card Should You Use?

Carlos needs living room furniture. He finds furniture he likes at a local department store for $1,500. He can afford to pay $200 per month on furniture.

Carlos can get a credit card from the store with no finance charge for the first 6 months. The minimum monthly payment is 2% of the unpaid balance. Starting with the seventh month, the annual percentage rate is 20%.

Carlos recently opened a new bank credit card account. It has an annual percentage rate of 12%. The minimum monthly payment is 4% of the unpaid balance.

Answer each question. Round to the nearest percent.

1. Suppose Carlos uses the store credit card to buy the set of living room furniture. What will the minimum payment be for the first month? How much of that is interest?

2. Suppose Carlos uses the bank credit card to buy the set of living room furniture. What will be the minimum payment for the first month? How much of that is interest?

3. Which card should Carlos use if he could pay the entire amount he owes in 8 months? Explain.

You Decide

Carlos has $1,200 in savings. How can Carlos reduce the amount of interest he pays on this purchase? Think about the APR on each card, the monthly payments he could make, and the money he already has in savings.

Summary

There are good and bad things about having a credit card.
To open a credit card account, you must complete an application.
You need to protect your credit card as if it were cash.
To find a monthly finance charge, multiply the balance by the monthly percentage rate.
A monthly statement shows information about your credit card account.
To find the new balance on a credit card account, add the new charges to the unpaid balance and subtract the credits.
If your total monthly credit card payments equal more than 20% of your monthly take-home pay, you should stop buying on credit.
The annual percentage rate for a cash advance is often more than the annual percentage rate for a purchase.

available credit

co-applicant

credit card

credit limit

finance charge

minimum payment

Vocabulary Review

Complete the sentences with the words from the box.

1. A person who applies with you for a credit card account is a _____.

2. The _____ is the most money you can borrow on a credit card.

3. The difference between your credit limit and the balance on your account is your _____.

4. A fee you pay each month for borrowing money is the _____.

5. A _____ allows you to buy items now and pay for them later.

6. The least amount of your balance that you must pay to the credit card company each month is a _____.

Chapter Quiz

Solve each problem. Show your work.

1. You have worked for your current employer for 60 months. On a credit card application, you have to write how many years you have worked at your present job. What will you write?

2. Julia's friend wants to use Julia's credit card to buy some clothes. Should Julia give her card to her friend? Explain.

3. Francis has an unpaid balance of $480 on his credit card. The annual percentage rate is 12%. What is the finance charge on his next monthly statement?

4. Joan has a credit limit of $900. Her new balance is $450. What is Joan's available credit?

5. Your monthly take-home pay is $900. Your monthly credit card payments are about $135. What percent of your take-home pay is used for your credit card payments?

6. The annual percentage rate for a cash advance is 24%. What is the total finance charge for one month on a $600 cash advance?

Maintaining Skills

Write each decimal as a percent.

1. 0.92

2. 0.5

3. 0.295

Compute.

4. 670 + 350

5. 900.6 + 400.0

6. 1,000 + 850

7. 24 − 1.2

8. 3.6 ÷ 18

9. 45 × 0.15

Bank loans allow you to pay for expensive things, such as college tuition, a car, or a home. Why would you need a loan?

Learning Objectives

LIFE SKILLS

- Learn about different types of loans and determine the best rate.

- Learn about steps you can take to maintain a good credit rating and improve a bad credit rating.

- Find the total amount paid on a loan and the monthly payment.

- Complete a loan application.

- Learn about loan fees.

MATH SKILLS

- Add, subtract, multiply, and divide with money.

- Compare percents.

- Find the percent of a number.

Words to Know

loan	to let someone borrow something, such as money; money that is borrowed
lending institution	a business or organization that loans money to people
mortgage	a loan used to buy a home
term	a set amount of time
qualify	to show you are able to take on a responsibility, job, or task, such as repaying a loan
lien	a claim on property or other things you own, which guarantees your payment of debt
cosigner	someone who signs a loan contract with a borrower and promises to pay back the loan if the borrower doesn't pay
collateral	something of value used as security for a loan

Project: Choosing a Personal Loan

Visit two banks in person or on the Internet. Find out about personal loans.

- What can you use a personal loan for?
- What is the interest rate?
- How much can you borrow?
- How do you pay back the loan?

Write a report. Include a bar graph that compares the interest rates and the maximum amounts you can borrow.

Someday you may need a **loan.** You may need extra money to pay for a house or a car. You can go to a **lending institution**, such as a bank or credit union, to get a loan.

Some different types of loans are a student loan for school bills, a car loan, and a **mortgage** for a house. Another type of loan is a personal loan for extra money to pay for unexpected expenses such as medical bills.

Nicole did some research on different loans. The chart below shows what she found in her area.

Type of Loan	Term of Loan	Interest Rate
Student loan	7 to 15 years	5% to 16%
New car loan	2 to 6 years	4% to 8%
Mortgage	15 to 30 years	5% to 9%
Personal loan	15 days to 4 years	8% to 20%

The **term** of a loan is the time you have to pay back the loan. The interest is the amount of money the bank charges you to borrow the money. The amount that you have to pay back will include the amount you borrowed plus the interest. Shop around for a loan, to find the lowest interest rate.

▶ **EXAMPLE**

Did You Know?
You want the highest interest rate on a savings account. You want the lowest interest rate on a loan.

Nicole went to a bank and a finance company to find the best personal loan. The bank offers an 8.5% interest rate. The finance company offers an 8.56% interest rate. If both lending institutions have the same charges, where should Nicole apply for the personal loan?

Compare the interest rates.

bank   finance company
8.5%   <   8.56%

Nicole will pay less interest at the bank. She should apply for her loan at the bank.

Practice and Apply

Solve each problem. Use the chart on page 156. Show your work.

1. What different types of loans can you get?

2. Suppose you need a student loan. What could be the term of the loan? What is the lowest interest rate you could pay?

3. James wants a car loan. What is the highest interest rate he could be charged for the loan?

4. Lori wants a mortgage for a house. Lori is young and wants as much time as possible to pay off the loan. What is the longest amount of time she could have to pay off a mortgage? (Hint: The term of a mortgage is the length of time given to pay it off.)

5. Zack wants to buy a car. He compares two car loans. The interest rate on one loan is 4.875%. The interest rate on the other loan is 4.87%. Which rate is better? Why?

6. Suppose you want to buy a car. You compare two different loans. The interest rate on one is 4.75%. The interest rate on the other is 4.8%. What is the difference between the two rates? (Hint: Subtract.)

7. **IN YOUR WORLD** Suppose you need a loan. Describe the type of loan you need. Tell why you need the loan. Decide how much money you need to borrow, the term of the loan, and the interest rate.

Suppose you need a loan from a bank. Do you **qualify** for the loan? To get a loan, it is important for you to have a good employment history and a good credit rating. This means that you can be trusted to pay back the loan. To get a good credit rating, you will need:

- A steady job

- A credit history that shows you pay your bills on time

- A checking or savings account

There are some things that make it harder to qualify for a loan. Ask yourself these questions:

- Is my address only temporary (for example, in a hotel or in someone else's home)?

- Has a **lien** ever been placed on anything I own?

- Have any of my bills been turned over to a collection agency because they were not paid on time?

- Am I under 21? (If yes, you may need a **cosigner**.)

If you have a bad credit rating, you can change it. Here are some things you should do:

- Pay every one of your bills on time.

- Use **collateral** to get an installment loan. Then pay the loan in full, making payments on time.

- Get an account at a department store. Charge only in small amounts that you pay promptly.

When you apply for a loan, you will need identification such as a driver's license, Social Security card, or birth certificate. You will also need your last earnings statement and a list of people who will say that you are responsible. These people are your references.

Did You Know?
A *lien* gives someone the right to take your property if you don't repay a debt.

Here's a Tip!
It's risky to be a cosigner. Don't do it unless you can trust the person to pay back the loan. Otherwise, you will have to pay back the loan.

Practice and Apply

Write the letter of the correct answer.

1. What do you need to get a loan?
   A. a steady job
   B. a temporary address
   C. overdue bills

2. What might make it harder for you to get a loan?
   A. having a checking account
   B. getting good grades
   C. making late payments on your bills

3. What can you do to help change a bad credit rating?
   A. Move.
   B. Quit your job.
   C. Pay every one of your bills on time.

4. **WRITE ABOUT IT** Suppose you have a lot of credit card debt. Describe the steps you will take to pay off the debt.

Maintaining Skills

Find the percent of each number.

1. 10% of 80	**2.** 15% of 200	**3.** 45% of 115
4. 75% of $800	**5.** 7% of $2,500	**6.** 18% of $5,600
7. 30% of $900	**8.** 5% of $32	**9.** 16% of $70

What Will the Loan Cost?

When you borrow money, you pay interest. The amount of interest depends on how much you borrow, the interest rate, and the term of the loan.

▶ **EXAMPLE 1**

Nicole took out a student loan for $1,500 at 19% simple interest for 2 years. How much was the total amount she paid on the loan?

STEP 1 Change the interest rate to a decimal.

19% = 0.19

STEP 2 Multiply the principal of the loan by the decimal by the time in years.

$1,500 × 0.19 × 2 = $570

STEP 3 Add the interest to the principal.

$1,500 + $570 = $2,070

The total amount Nicole paid on the loan was $2,070.

▶ **EXAMPLE 2**

Aaron bought a used car for $6,300. He took out a loan at 12% simple interest for 3 years. How much were the monthly payments?

STEP 1 Change the interest rate to a decimal.

12% = 0.12

STEP 2 Multiply the amount of the loan by the decimal by the time in years.

$6,300 × 0.12 × 3 = $2,268

STEP 3 Add the interest to the principal.

$6,300 + $2,268 = $8,568

Here's a Tip!
3 years is 36 months.

STEP 4 Divide the total by the time in months.

$8,568 ÷ 36 = $238

The monthly payments were $238.

Practice and Apply

 Find the simple interest on each loan. Then find the total amount that will be paid on each loan. Use a calculator if you like.

	Principal	Interest Rate	Term of the Loan	Interest	Total Amount Paid
1.	$900	20%	1 year	?	?
2.	$2,500	19%	2 years	?	?
3.	$1,500	18%	2 years	?	?
4.	$3,000	21%	4 years	?	?
5.	$4,500	15%	3 years	?	?
6.	$2,480	12%	5 years	?	?

Find the number of monthly payments for each loan. Then find the monthly payment. Use a calculator if you like. (Hint: There are 12 months in a year.)

	Total Amount Paid	Term of the Loan	Number of Monthly Payments	Monthly Payment
7.	$1,080	1 year	?	?
8.	$2,520	3 years	?	?
9.	$3,000	5 years	?	?
10.	$6,000	2 years	?	?
11.	$8,448	4 years	?	?

12. You have a two-year personal loan. Each month you pay $102. What is the total amount you pay on the loan?

13. **CRITICAL THINKING** Marcel bought a used car for $5,000. He took out a loan at 7% simple interest for 2 years. How much was each monthly payment?

8·4 ▶ What's on a Loan Application?

You decide to apply for a loan. Now you must fill out a loan application. Read each section of the application.

Loan Application

SECTION A: YOUR CREDIT REQUEST

WHAT DO YOU WANT THIS LOAN FOR?_____

LOAN $ _____FOR ____MONTHS ☐ FIXED INTEREST RATE   (Explain Purpose)

THIS APPLICATION IS ☐ IN YOUR NAME ALONE ☐ JOINTLY WITH _____
If applying jointly, each of you must complete a separate application.

SECTION B: YOURSELF

FIRST NAME	MIDDLE INITIAL	LAST NAME	

CURRENT ADDRESS	STREET	APT. NO.	TIME THERE YRS. MOS.

CITY	STATE	ZIP CODE	☐OWN ☐RENT

PREVIOUS ADDRESS   STREET   APT. NO.   TIME THERE
(If at current address less than three years)   YRS.   MOS.

CITY	STATE	ZIP CODE	DATE OF BIRTH

SOCIAL SECURITY NO.	DRIVER'S LICENSE NO.

HOME PHONE NO.	DEPENDENTS OTHER THAN SELF OR SPOUSE

BEST TIME TO CALL YOU   BEST PLACE TO CALL YOU
☐MORNING ☐AFTERNOON ☐EVENING   ☐HOME ☐WORK

MARITAL STATUS   ☐MARRIED   ☐SINGLE   ☐SEPARATED   ☐DIVORCED

SECTION C: YOUR EMPLOYMENT

CURRENT EMPLOYER	WORK PHONE NO.

EMPLOYER'S ADDRESS	CITY	STATE	ZIP CODE

TIME THERE   OCCUPATION
☐YRS. ☐MOS.

PREVIOUS EMPLOYER (if with current employer   OCCUPATION   TIME THERE
                 less than three years)   ☐YRS. ☐MOS.

EMPLOYER'S ADDRESS	CITY	STATE	ZIP CODE

SECTION D: YOUR INCOME	
MONTHLY GROSS SALARY/WAGES	$
DIVIDENDS AND INTEREST	$
OTHER INCOME (DESCRIBE)	$
YOUR TOTAL MONTHLY INCOME	$

SECTION E: YOUR BANKING RELATIONSHIPS			
TYPE OF ACCOUNT FINANCIAL INSTITUTION		ACCOUNT #	BALANCE
CHECKING			
SAVINGS			
OTHER (SPECIFY)			

Practice and Apply

Answer each question. Use the loan application on pages 162–163.

1. Look at Section A. Gloria wants to borrow $2,400 for 2 years. What will she write on the line before the word MONTHS?

2. Look at Section B. Gloria has lived at her current address for 30 months. Should she fill out any information under PREVIOUS ADDRESS? Explain.

3. Look at Section C. Gloria has worked at a cycle shop for $3\frac{1}{2}$ years. What will she put in the boxes on the third line?

4. Before she took a job at the cycle shop over 3 years ago, Gloria worked as a cashier in a department store. Will she include this information on the application? Explain.

5. Gloria has a checking account at Valley Bank. Her checking account number is 1182-6609. Her balance is $63.04. How will she fill out the first line in Section E?

6. **WRITE ABOUT IT** Write a letter to a bank, asking for a loan application.

What Is the Fee on a Loan?

When you take out a loan, you may need to pay a loan fee. You can have it taken out of the amount you borrow or you can pay it with money you already have.

▶ **EXAMPLE 1**

Aaron took out a loan for $1,200. The fee of $55 was deducted from the loan. How much was left?

Subtract the fee from the loan.

$$\begin{array}{r} \$1,200 \\ -\quad 55 \\ \hline \$1,145 \end{array}$$

$1,145 was left from the loan.

Sometimes the fee is calculated as a percent of the loan.

▶ **EXAMPLE 2**

Lee took out a loan for $1,800. The fee was 4% of the loan and was deducted from the loan. How much was left?

STEP 1 Change the percent to a decimal.

$$4\% = 0.04$$

STEP 2 Find the fee. Multiply the loan by the rate.

$$\begin{array}{r} \$1,800 \\ \times\quad 0.04 \\ \hline \$72.00 \end{array}$$

STEP 3 Subtract the fee from the loan.

$$\begin{array}{r} \$1,800 \\ -\quad 72 \\ \hline \$1,728 \end{array}$$

$1,728 was left from the loan.

After you receive your loan, you must make a loan payment every month for the term of the loan. If you cannot make a payment on time, you should

• Contact the bank before the payment is due.

• Explain that you will make the payment late.

Here's a Tip!
Ask the bank to increase the term of the loan if you continue to have trouble making the monthly payments. This will increase the cost of the loan but also decrease the amount you must pay each month.

Practice and Apply

Solve each problem. Show your work.

	Amount of Loan	Deducted Fee	Amount Left
1.	$2,300	$175	?
2.	$1,500	$75	?
3.	$2,500	$150	?
4.	$3,500	$180	?

5. Morris took out a loan for $2,500. The loan fee was 5% of the loan. The fee was deducted from the loan. How much of the loan was left?

6. Ebony took out a loan for $900. The loan fee was 3% of the loan. The fee was deducted from the loan. How much of the loan was left?

7. Jeremy took out a loan for $4,000. The loan fee was 6% of the loan. The fee was deducted from the loan. How much of the loan was left?

8. Suppose you could not go to work because you were sick for the first two weeks of November. You are only paid for 3 sick days. You have a loan payment due on November 15th. You cannot make the loan payment until your next payday on November 20th. What should you do?

9. You find you cannot make your loan payments every month because your other expenses are too high. What should you do?

10. **WRITE ABOUT IT** Suppose you have a loan, and have just lost your job. You find another job. However, you will not be able to make a payment on your loan for two months. Write a letter to your bank, explaining your situation and asking for more time.

8·6 ▶ Problem Solving

Solve each problem. Show your work.

1. Mark took out a car loan for $22,500. He will pay 11% simple interest for 4 years. How much interest will he pay?

2. Joe took out a loan for $1,200. He will pay 15% simple interest. He plans to repay the loan in one year by working extra hours. His net earnings are $20 an hour. How many extra hours will Joe need to work to pay the loan plus interest? (Hint: Divide the total amount he will pay by $20.)

3. A farmer needs a new truck. A three-year loan costs $210 a month. Her net income is $19,000 per year. She can only afford to spend 15% of her net income on the loan. Can she afford the loan? Explain.

4. **OPEN ENDED** Nicole took out two different loans. Each loan was for $1,000. The term of each loan was 1 year. Altogether, the simple interest on both loans was $100. What could the annual percentage rate have been for each loan.

Calculator

You can use a calculator to find the interest on a loan.
Then add the amount of the loan to find the total amount paid.

Find the total amount paid on a loan for $5,000 at 12% simple interest for 3 years.

Press: ⑤ ⓪ ⓪ ⓪ ✕ ① ② % ✕ ③ ＋ ⑤ ⓪ ⓪ ⓪ ＝ | 6,800 |

Find the total amount paid on each loan.

	Loan Amount	Annual Rate	Term	Total Amount Paid
1.	$8,000	7%	2 years	?
2.	$12,000	18%	4 years	?
3.	$13,000	5%	3 years	?

ON-THE-JOB MATH:
Bank Loan Representative

Daniel is a bank loan representative. He helps people apply for loans by phone. He checks each customer's credit and employment history and gives each customer a rating. The better the rating, the lower the interest rate.

The chart below shows interest rates for each credit rating. It also shows different terms for a loan and the percent of the loan needed as a down payment.

New Car Loans				
Credit Rating	36 Months	48 Months	60 Months	Percent Down
Excellent	4.75%	5.19%	5.49%	0%
Good	5.39%	5.42%	5.49%	5%
Average	6.29%	6.29%	6.29%	5%
Fair	7.99%	7.99%	7.99%	10%
Poor	13.95%	18.59%	20.95%	10%

1. A customer has a credit rating of "Average" and wants a 48-month car loan. What would the interest rate be? What percent of the cost of the car would be needed for a down payment?

2. What percentage would a customer save if Daniel gave him a "Good" rating instead of "Average" for a 60-month loan?

3. How much would someone with a "Fair" credit rating need as a down payment for a $20,000 loan?

You Decide

A customer with a poor credit rating is buying a new car. She can pay a down payment of $1,000. Can she get a $10,000 loan? How long would you recommend she take to repay the loan? Explain the benefits of taking a 36-month loan instead of a 60-month loan.

Summary

Some different types of loans are student loans, car loans, mortgages, and personal loans.

You can find the best interest rate by comparing what different banks offer.

One way to maintain a good credit rating is to pay your bills on time. To improve a bad credit rating, you can pay off all your bills.

The total amount you will pay on a loan is equal to the principal plus the interest.

To find the monthly payment, divide the total amount you will pay on the loan by the number of months in the term of the loan.

The first step you take to get a loan is to complete a loan application.

Sometimes you need to pay a fee to get a loan. The fee may be deducted from your loan.

collateral

cosigner

lending institution

lien

loan

mortgage

qualify

term

Vocabulary Review

Complete the sentences with words from the box.

1. Money that is borrowed is a _____ .

2. To show that you are able to take on a responsibility, job, or task means that you _____ .

3. _____ is something of value used as security for a loan.

4. A _____ signs a loan contract with a borrower and promises to pay back the loan if the borrower doesn't pay.

5. A _____ is a claim on property or other things you own, which guarantees payment of your debt.

6. A business or organization that loans money to people is a _____ .

7. A set amount of time is a _____ .

8. A _____ is a loan used to buy a home.

Chapter Quiz

Solve each problem. Show your work.

1. Name three types of loans you can get from a bank or other lending institution.

2. Sam wants to buy a car. One loan has an interest rate of 6.29%. Another loan has a rate of 6.3%. Which rate is better? Explain.

3. What is one thing you can do to improve your credit rating?

4. Suppose you get a personal loan for $1,800 with an interest rate of 18%. The loan is for 3 years. What is the total amount you pay on the loan, including interest?

5. Suppose you get a personal loan for $2,400 with an interest rate of 12%. The loan is for 4 years. What is the monthly payment for the loan?

6. Claudia is filling out a loan application. She moved to a new apartment 4 months ago. What sections will she need to fill in on Section B of the application on page 162?

7. Jessica took out a loan for $800. The loan fee was $40. The fee was deducted from the loan. What was left from the loan?

8. Mustafa took out a loan for $6,500. The loan fee of 5% was deducted from the loan. What was left from the loan?

Maintaining Skills

Divide to find the percent.

1. $\frac{3}{4}$
2. $\frac{17}{50}$
3. $\frac{27}{1,000}$

Find the percent of each number. Round to the nearest percent.

4. $144 out of $400
5. $386 out of $500
6. $18 out of $207

Unit 4 **Review**

Write the letter of the correct answer. Use the bar graph to answer questions 1 and 2.

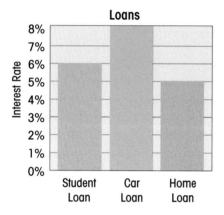

Loans

1. What is the interest rate on a home loan?

   A. 5%

   B. 6%

   C. 7%

   D. 8%

2. What is the difference between the interest rate on a student loan and the interest rate on a car loan?

   A. 1%

   B. 2%

   C. 3%

   D. 4%

3. Liam's credit card bill is $325. The interest rate is 18%. Liam pays $25. What is the interest on the remaining balance for 1 month?

   A. $4.50

   B. $4.88

   C. $45.00

   D. $54.00

4. The previous balance on Rita's credit card is $180. She charges $150. Rita makes a payment of $40. What is her unpaid balance, not including the finance charge?

   A. $10

   B. $70

   C. $290

   D. $370

5. Kevin has a student loan for $8,000 at 6% simple interest for 7 years. How much will he pay in interest?

   A. $336

   B. $4,640

   C. $11,360

   D. $3,360

6. Marta took out a personal loan for $1,500 at 13% simple interest. How much will she have paid in total if she repays the loan in 2 years?

   A. $390

   B. $1,890

   C. $1,980

   D. $1,110

Challenge

You have a car loan for $5,000 at 7% interest for 4 years and a personal loan for $5,000 at 11% interest for 3 years. For which loan will you pay more interest? How much more?

Some people live in an apartment. Some people live in a house. The cost of rent and utilities may depend on the housing you choose.

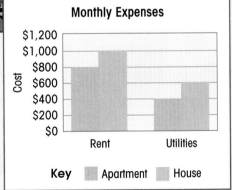

Monthly Expenses

The bar graph shows some typical expenses for renting an apartment and renting a house. Use the graph to answer each question.

1. What is the monthly rent for the apartment?

2. What is the monthly rent for the house?

3. Is the cost of utilities greater for the apartment or for the house?

171

Chapter 9

Finding a Place to Live

Finding a place to live may be easier if you share your apartment. Then you can share the expenses, such as rent. How much do you think you can afford to spend for rent?

Learning Objectives

LIFE SKILLS

- Determine how much you can afford for rent.
- Identify the best type of place for you to rent.
- Understand what a lease and a security deposit are and how they work.
- Determine if you can afford to buy a house.
- Calculate the down payment for a house.
- Calculate mortgage payments.
- Calculate the closing costs, real estate tax, and homeowner's insurance.

MATH SKILLS

- Add, subtract, multiply, and divide with money.
- Find the percent of a number.

Words to Know

landlord	a person who owns an apartment or house and rents it to others
rent	the monthly fee paid to a landlord for a place to live
abbreviation	a shortened form of a word
lease	a legal contract between a renter and the landlord
security deposit	the money a renter pays to the landlord before moving in; the money will be returned when the renter moves out if the place is clean and has not been damaged
down payment	the money paid in advance as part of the purchase of a house
closing costs	the fees that you pay when you take ownership of a house
real estate tax	a local tax on property that is paid by the owner
homeowner's insurance	insurance that will pay for damage or loss caused by fire, smoke, theft, or some types of weather

Project: What About Location?

You need to find a place to live that is close to school or work, public transportation, and a grocery store. Ask yourself these questions:

- How many miles away from my job can I live?
- How much will daily transportation cost?
- Are there stores and shops nearby?

Find a map or create a map on a travel Web site such as www.mapquest.com. Draw a circle on the map with your school or place of work at the center. The circle should have a radius of about 1 mile. Then decide if the area meets your needs.

Many people live in an apartment or a house they do not own. They pay a monthly fee to a **landlord,** or property owner. The fee is called **rent.**

Your rent should be about 25% of your gross monthly income, if possible. Paying more than 30% of your income for rent may make it hard to balance your budget.

▶ **EXAMPLE 1**

Chantal works as a receptionist. Her monthly income is $2,350. How much will Chantal pay for rent each month if she spends 25% of her income?

STEP 1   Change the percent to a decimal.
25% = 0.25

STEP 2   Find 25% of the monthly income.
$2,350 × 0.25 = $587.50

Chantal will pay $587.50 for rent each month.

▶ **EXAMPLE 2**

Tak is an electrician. He earns $28 an hour and works 40 hours a week. Tak found an apartment for $1,200 a month. Can he afford the apartment? He needs to spend less than 30% of his income on rent.

Here's a Tip!
You can estimate the monthly income if you know the weekly income. Think of a month as 4 weeks.

STEP 1   Find the weekly income.
$28 × 40 = $1,120

STEP 2   Find the monthly income. Multiply the weekly income by 4.
$1,120 × 4 = $4,480

STEP 3   Find 30% of the monthly income.
$4,480 × 0.30 = $1,344

STEP 4   Compare.
$1,344 > $1,200

Tak can afford to pay $1,200 a month for rent. This is less than 30% of his income.

Practice and Apply

Complete the chart. Each person spends 25% of his or her income on rent. (Hint: For Exercises 4–6, first find the monthly income.)

	Name	Income	Monthly Rent
1.	Rachel	$2,150 per month	?
2.	Durran	$800 per month	?
3.	Tim	$1,500 per month	?
4.	Nicole	$26,400 per year	?
5.	Juan	$28,800 per year	?
6.	Alice	$18,600 per year	?

Solve each problem. Show your work.

7. Erika's hourly wage is $10. She works 35 hours a week. What is her weekly income? What is her monthly income? If Erika pays 30% of her income for rent every month, what does she pay?

8. Kim's hourly wage is $14. She works 37 hours per week. What is her weekly income? What is her monthly income? Kim can pay at most 30% of her income on rent. What is the most she can pay for rent every month?

9. Erin makes $18 an hour. She works 40 hours a week. She found an apartment for $1,000 a month. Should Erin rent the apartment? Explain.

10. **CRITICAL THINKING** You found an apartment you really like. The rent is $625 per month. This is 25% of your monthly income. How much is your monthly income?

You decide to rent an apartment. You look in the classified section of your local newspaper and see the ads below. The ads tell you about rooms, apartments, and houses for rent in your area. Some ads may use an **abbreviation**. To find out what the abbreviations mean, look on page 366 of this book.

CALIFORNIA AVE. —Nice studio from $500, rent incls. ht. + h.w. 410-555-3434	**VIRGINIA ST.** —2BR fully renov., $550/mo+sec dep. Call 410-555-2332	**UTAH PL.** —extra large 1 BR, avail. immed. Plenty closet space. New appls, AC, gas, heat, $525+$525 SD. 410-555-9888

How do you decide which apartment to rent? First, list the things you want in an apartment.

My Apartment

1. The rent should be no more than $570 a month.
2. It should have one bedroom.
3. It should be close to work.

Then compare your list to the ads. Which apartment has the things you want?

- All of the apartments have rent that is less than $570 a month.

- The apartments on Virginia Street and Utah Place each have at least one bedroom.

- The apartment on Utah Place is close to work.

You choose the apartment on Utah Place. This apartment has all the things you want.

Practice and Apply

Solve each problem. Use the classified ads on the right and the abbreviations on page 366. Show your work.

1. Look at the ad for the apartment on First Street. What does the abbreviation AC mean?

2. Which apartment is the least expensive? Which one is the most expensive? What is the difference between the most expensive and the least expensive rent?

3. Look at the ad for the apartment on Second Avenue. How many bedrooms does the apartment have? Explain.

4. Mrs. Cooper wants a two-bedroom apartment so her daughter can have her own bedroom. She also wants a parking space. Her income is $2,960 a month. Which apartment should she choose?

5. Michael is a student. He wants an apartment that is close to campus. He can afford to spend $380 a month. Which apartment should he choose?

6. Sanji earns $2,000 a month. He wants an apartment with hardwood floors in the bedroom. He also wants to do his laundry at home. Which apartment should he choose?

7. Lenore needs an apartment immediately. She would like cable TV included. Her net income is $2,200 a month. Which apartment should she choose? Explain.

8. **IN YOUR WORLD** You plan to share an apartment with a friend to help you afford what you need. List the things you feel are important to do if you share an apartment.

FIRST ST. 2 BR $530/mo, AC, cable, hwd flrs, avail. immed. Call 614-555-8797

SECOND AVE. studio, hwd flrs, easy access to university, $360. Call 614-555-8777

THIRD PL. 2BR, AC, Cable assigned parking, $685, avail. immed. Call 614-555-3332

FOURTH DR. 1BR, 1 BA, hwd flrs, washer/dryer in basement, avail. in Oct. $425. Call 614-555-4444

What Is in a Lease?

Before you can move into your new apartment, you need to sign a **lease**. A lease is a legal contract between you and your landlord. A lease usually includes the following:

- a description of the apartment

- whether or not pets are allowed

- how much the monthly rent is and when it is due

- the amount of the **security deposit**, which covers any damage you cause

- the length of time you agree to stay in the apartment

- how much notice you must give the landlord before you move out

Here's a Tip!
Pay your rent on time to avoid paying a late fee.

Read the lease carefully before you sign it. If there are parts you disagree with, discuss them with the landlord. Once you sign the lease, you should obey its terms. If you move out early, you may still be responsible to pay the rent for the full term of the lease.

EXAMPLE

Gwen signed a one-year lease that began on January 1. The rent is $800 a month. She decided to move out at the end of October. How much will Gwen owe in rent when she moves out?

STEP 1 Find how many months are left on the lease.
From November to December is 2 months.

STEP 2 Multiply the rent by the number of months left on the lease.
$800 × 2 = $1,600

Gwen will owe $1,600 in rent when she moves out.

Sometimes a lease contains an *early out* clause. This clause allows you to leave early by paying a fee. This fee is paid instead of the remaining rent payments.

Practice and Apply

Solve each problem. Show your work.

1. Estella signed a one-year lease. She pays $720 per month in rent. If she wants to move out three months early, how much money will she owe when she moves out?

2. Rasheed signed a two-year lease. He pays $565 per month in rent. If he wants to move out six months early, how much money will he owe when he moves out?

3. Joshua pays $795 per month in rent. He signed a one-year lease to begin on January 1. He decided to move out at the end of June. How much money does Joshua owe when he moves out?

4. Alisha pays $550 per month in rent. She signed a one-year lease that began on September 1. She decided to move out at the end of December. How much money does Alisha owe when she moves out?

5. **CRITICAL THINKING** Karin pays $840 per month in rent. She signed an 18-month lease that began June 15. By September 14 of the next year, she had to move. How much did Karin have to pay when she moved out?

Maintaining Skills

Multiply or divide.

1. $18,000 ÷ 2.5   2. $4,000 × 3.5   3. $75,000 ÷ 2.5

4. $18,000 × 2.75   5. $10,000 × 4.5   6. $48,950 ÷ 5.5

What Is a Security Deposit?

You found an apartment you like and can afford. What happens next? Usually, you must pay the first month's rent when you sign the lease. You must also pay a security deposit. Then you can move in.

► **EXAMPLE 1**

Consumer Beware!
Be sure you set aside enough money if you are planning to move. The money you need for rent, a security deposit, and moving your belongings can add up.

Wanda found an apartment that rents for $860 per month. Before she moves in, she must pay the first month's rent. She also must pay the security deposit equal to one and a half months' rent. How much money does Wanda pay altogether before she can move in?

STEP 1 $1\frac{1}{2} = 1.5$ So multiply the rent by 1.5 to find the security deposit.

$860 × 1.5 = $1,290

STEP 2 Add the security deposit to the first month's rent.

$1,290 + $860 = $2,150

Wanda pays $2,150 before she can move in.

When you move out, your landlord returns your security deposit if you have followed the lease agreement. If you damaged the apartment, the landlord may keep some or all of the deposit to pay for repairs.

► **EXAMPLE 2**

When Wanda moved out of her apartment, there was damage to the walls. The pictures she had were hung on large nails that left holes in the walls. The landlord paid $580 to have the walls repaired and painted. Wanda had paid a security deposit of $1,290 when she moved in. How much money did she get back when she moved out?

Subtract the cost of the damage from the security deposit.

$1,290 – $580 = $710

Wanda got back $710 when she moved out.

Practice and Apply

Solve each problem. Show your work.

1. The apartment you like rents for $700 a month. The landlord requires two months' rent for the security deposit. How much is the security deposit?

2. Juno found an apartment that rents for $625 a month. The landlord requires one month's rent for the security deposit and the first month's rent before Juno can move in. How much money does Juno pay altogether before he can move into his new apartment?

3. Richard found an apartment that rents for $485 a month. The landlord requires one and a half months' rent for the security deposit and the first month's rent before Richard can move in. How much money does he pay altogether before he can move into his new apartment? Use a calculator if you like.

4. Emil moved out of his apartment. His dog had damaged the rugs. The landlord paid $400 to have the rugs cleaned. Emil had paid $1,100 as a security deposit. How much did he get back when he moved out?

5. Natasha moved out of her apartment. There were stains on the carpet that could not be cleaned. The landlord paid $600 to have the carpet replaced. Natasha had paid $900 as a security deposit. How much did she get back when she moved out?

6. **WRITE ABOUT IT** Brian moved into an apartment last year. He paid a security deposit of $900. When the lease was up, he moved out. He took photographs of the apartment the day he moved out. These photographs showed that there was no damage to the apartment. The landlord did not return the security deposit to Brian. Write a letter that Brian could send to the landlord, explaining why the security deposit should be returned.

Can You Afford to Buy a House?

You have been living on your own for several years. You have a steady job and you rent your own apartment. You have money in your savings account. Now you want to buy a house.

Buying a house can be very expensive. Most people need to take out a loan to buy a house. This loan is called a mortgage. Banks and other financial institutions offer mortgages. There are different types of mortgages. You will want to find the best mortgage for you.

To find out how much money you can afford to borrow, you can use the *banker's rule*. The banker's rule says that you can afford to borrow up to 2.5 times your annual income.

Did You Know?
You may not be able to get a mortgage from a finance company. But you may be able to get an FHA mortgage. See www.FHA.com

▶ **EXAMPLE 1**

Martin wants to buy a house. He earns $25,600 a year. How much can he afford to borrow to buy a house?

Multiply Martin's annual income **by 2.5.**

$25,600 × 2.5 = $64,000

Martin can afford to borrow $64,000 to buy a house.

Suppose you find a house you like. How do you know if you can afford it? You can use the banker's rule to find the monthly income you need.

▶ **EXAMPLE 2**

You found a house you like. You need to borrow $72,000 to buy the house. What annual income do you need to afford to borrow $72,000?

Divide the amount you need to borrow **by 2.5.**

$72,000 ÷ 2.5 = $28,800

You need an annual income of $28,800 to afford to borrow $72,000.

Practice and Apply

Find the amount of money each person can afford to borrow. Use a calculator if you like. (Hint: Use the banker's rule.)

	Name	Annual Income	Amount That Can Be Borrowed
1.	Greta	$34,000	?
2.	Michael	$27,500	?
3.	Salvador	$30,000	?
4.	Amber	$32,300	?
5.	Lee	$24,700	?

Solve each problem. Show your work. Use a calculator if you like.

6. Cecilia found a house she likes. She needs to borrow $95,000 to buy the house. What annual income does Cecilia need to afford to borrow the money?

7. Ryan found a house he likes. He needs to borrow $55,650 to buy the house. What annual income does Ryan need to afford to borrow the money?

8. Juan needs to borrow $82,000 to buy the house he likes. What annual income does Juan need to afford to borrow the money?

9. Theresa earns $10 an hour. What is Theresa's annual income? How much can Theresa afford to borrow for a house? (Hint: There are 2,080 working hours in a year.)

10. **WRITE ABOUT IT** How does the banker's rule prevent people from borrowing more money than they can afford to pay back?

You will probably need to make a **down payment** before you can get a mortgage. A down payment is part of the money used to buy a house. The down payment a lender requires can be as much as 20% of the price of the house.

▶ **EXAMPLE 1**

Sheila found a house for $154,900. She needs a 20% down payment. How much does Sheila need for the down payment?

Multiply the price of the house by the rate of the down payment. 20% = 0.20	$154,900 $\times$ 0.20 $30,980

Sheila needs $30,980 for the down payment.

The down payment reduces the amount of money you will borrow to buy a house. Once you know the amount of the down payment, you can find out how much money you will need to borrow. The amount you borrow is called the mortgage.

▶ **EXAMPLE 2**

Larry found a house for $110,000. He pays 20% of the price of the house as a down payment. How much will the mortgage be?

STEP 1   Multiply the price of the house by the rate of the down payment.

$110,000
$\times$ 0.20
$22,000

STEP 2   Subtract the down payment from the price of the house.

$110,000
$-$ 22,000
$88,000

Did You Know?
Fannie Mae and Freddie Mac are abbreviations for two mortgage companies that were created by Congress. They make housing more affordable for many people.

The mortgage will be $88,000.

Some mortgage lenders, such as Fannie Mae and Freddie Mac, require only a small down payment. It may be from 3% to 5% of the price of the house.

Practice and Apply

Solve each problem. Show your work.

1. Justin found a house for $112,000. He needs a 20% down payment. How much money does Justin need for the down payment?

2. Grace found a house for $98,000. She needs a 5% down payment. How much money does Grace need for the down payment?

3. Stan found a house for $124,000. The bank requires 20% of the price of the house for the down payment. How much will the mortgage be?

4. Helene found a house for $85,000. She will pay 4% of the price of the house as a down payment. How much will the mortgage be?

5. Ben found a house for $104,000. He will pay 12% of the price of the house as a down payment. How much will the mortgage be?

 Find each down payment and mortgage. Use a calculator if you like.

	Name	Price of House	Rate of Down Payment	Down Payment	Mortgage
6.	Juan	$162,800	20%	?	?
7.	Bobby	$87,000	3%	?	?
8.	Susan	$132,000	5%	?	?
9.	Cheng	$75,000	11%	?	?

10. **CRITICAL THINKING** Charlie found a house for $120,000. He had $30,000 in a savings account to give to the bank as a down payment. What rate of down payment did Charlie pay? (Hint: Divide the down payment by the price of the house.)

How Much Is the Monthly Mortgage Payment?

You make payments on a mortgage every month for a number of years. Each payment includes interest on the loan. One type of mortgage is a fixed-rate mortgage. You pay the same interest rate for the entire term of the loan. The chart shows the monthly payments on a fixed-rate 30-year mortgage for different interest rates.

Monthly Payments for Fixed-Rate 30-Year Mortgage								
Mortgage Amount	Interest Rate							
	6%	6.5%	7%	7.5%	8%	8.5%	9%	10%
$40,000	$246	$252	$266	$280	$294	$308	$322	$351
$50,000	308	316	333	350	367	385	402	439
$60,000	369	379	399	420	440	461	483	527
$70,000	431	442	466	489	514	538	563	614
$80,000	493	506	532	559	587	615	644	702
$90,000	554	569	599	629	660	692	724	790
$100,000	616	632	665	699	734	768	805	878

EXAMPLE

Suppose you took out a 30-year mortgage for $80,000 at 8.5%. How much is the monthly mortgage payment? How much will you pay over 30 years?

STEP 1 Find the row for $80,000 and the column for 8.5%. Where the row and column meet is the monthly payment.

The monthly mortgage payment is $615.

STEP 2 Multiply the monthly mortgage payment by 12 to find how much you pay in a year.
$615 × 12 = $7,380

STEP 3 Multiply the amount you pay in a year by 30 to find how much you pay over 30 years.
$7,380 × 30 = $221,400

You will pay $221,400 over 30 years.

Practice and Apply

 Solve each problem. Use the chart on page 186. Show your work. Use a calculator if you like.

1. Nicole took out a 30-year mortgage for $90,000 at 7.5%. How much is her monthly mortgage payment?

2. Zack took out a 30-year mortgage for $70,000 at 9%. How much is his monthly mortgage payment?

3. Aaron took out a 30-year mortgage for $100,000 at 7%. How much will he pay over one year? (Hint: Multiply the monthly mortgage payment by 12.)

4. Yolanda took out a 30-year mortgage for $80,000 at 10%. How much will she pay over one year?

5. Manny took out a 30-year mortgage for $60,000 at 7.5%. How much will he pay over one year?

6. Kyle took out a 30-year mortgage for $40,000 at 9%. How much will he pay over 30 years? (Hint: First find how much he will pay in a year. Then multiply by 30.)

7. Gloria took out a 30-year mortgage for $70,000 at 7.5%. How much will she pay over 30 years?

8. Maura took out a 30-year mortgage for $100,000 at 8%. How much will she pay over 30 years?

9. CRITICAL THINKING Richard took out a 30-year mortgage for $90,000 at 6%. How much interest will he pay over 30 years? (Hint: First find how much he will pay over 30 years. Then subtract the amount of the mortgage.)

What About Closing Costs?

Before you can finalize the purchase of your new home, you will have to attend the closing. At the closing, ownership of the home is transferred to the buyer. There are several fees that you will have to pay.

These fees, called **closing costs,** must be paid at the closing. The closing costs are usually from 3% to 6% of the price of the house.

EXAMPLE 1

Wordwise
The *minimum* is the least. The *maximum* is the greatest.

Karla is buying a house for $130,000. The closing is this week. Karla wants to know about how much money she will need for closing costs. What is the minimum amount of money Karla will need to pay for closing costs? (Hint: The minimum amount is usually 3% of the price of the house.)

STEP 1 Change the percent to a decimal.   3% = 0.03

STEP 2 Multiply the price of the house by the closing cost rate.

$$\begin{array}{r} \$130,000 \\ \times \quad 0.03 \\ \hline \$3,900 \end{array}$$

The minimum amount Karla will need is $3,900.

EXAMPLE 2

Bryan is buying a $110,000 house. He wants to know about how much money he will need for closing costs. What is the maximum amount of money Bryan will need to pay for closing costs? (Hint: The maximum amount is usually 6% of the price of the house.)

STEP 1 Change the percent to a decimal.   6% = 0.06

STEP 2 Multiply the price of the house by the closing cost rate.

$$\begin{array}{r} \$110,000 \\ \times \quad 0.06 \\ \hline \$6,600 \end{array}$$

The maximum amount Bryan will need is $6,600.

Practice and Apply

Solve each problem. Show your work. Use a calculator if you like.

1. Jorge's closing costs are 3% of the price of the house. His new house costs $118,000. How much will Jorge have to pay for closing costs?

2. Roberta's closing costs are 6% of the price of the house. Her new house costs $172,000. How much will Roberta have to pay for closing costs?

3. Alia's closing costs are 5% of the price of the house. Her new house costs $120,000. How much will Alia have to pay for closing costs?

4. Orlin's closing costs are 4.5% of the price of the house. His new house costs $100,000. How much will Orlin have to pay for closing costs?

5. Patrick is buying a house for $85,000. He wants to know how much he will pay for closing costs. What is the minimum amount Patrick will need? (Hint: The minimum amount is 3% of the price of the house.)

6. Jada is buying a house for $142,000. She wants to know how much money she will need for closing costs. What is the maximum amount Jada will need? (Hint: The maximum amount is 6% of the price of the house.)

7. **CRITICAL THINKING** Armand wants to know how much he should expect to pay for closing costs on his new house. The price of his new house is $105,750. Armand knows he can expect to pay from 3% to 6% of the cost of the house. What is the minimum amount Armand should expect to pay for closing costs? What is the maximum amount Armand should expect to pay for closing costs?

9·9 ▶ What Is Real Estate Tax?

Some local governments raise money for schools and other services by taxing homeowners. If you own a house, you must pay **real estate tax.** Many homeowners pay part of their real estate tax each month or several times a year.

▶ **EXAMPLE 1**

Aaron pays $900 a year for the real estate tax on his house. He makes a real estate tax payment quarterly. How much does he pay each quarter for real estate tax?

Wordwise:
Quarterly means every 3 months, or 4 times a year.

Divide the real estate tax by 4 to find the amount of the quarterly payment.

$900 ÷ 4 = $225

Aaron pays $225 quarterly for real estate tax.

You can include a monthly real estate tax payment with your monthly mortgage payment. Then you only have to make one payment to the bank each month. The bank sends the tax payment to the government.

▶ **EXAMPLE 2**

Sheena pays $1,080 a year for the real estate tax on her house. She sends a tax payment each month to the bank along with her mortgage payment. Her mortgage payment is $615 a month. How much does she pay the bank each month?

STEP 1   Divide the real estate tax by 12 to find the monthly tax payment.

$1,080 ÷ 12 = $90

STEP 2   Add the monthly tax payment to the monthly mortgage payment.

$615 + $90 = $705

Sheena pays the bank $705 each month.

Practice and Apply

Solve each problem. Show your work.

1. Pete pays $1,200 a year for real estate tax on his house. He makes a real estate tax payment quarterly. How much does he pay each quarter for real estate tax?

2. Maura pays $912 a year for real estate tax on her house. She makes a real estate tax payment each month. How much does she pay each month for real estate tax?

3. Hannah pays $780 a year for real estate tax on her house. She makes a real estate tax payment each quarter. How much does she pay each quarter for real estate tax?

4. Chris pays $1,140 a year for real estate tax on his house. He sends a real estate tax payment each month to the bank with his mortgage payment. His mortgage payment is $720 a month. How much does he pay the bank altogether each month?

Each person in the chart below makes a monthly real estate tax payment with the mortgage payment. Complete the chart. Use a calculator if you like.

	Name	Annual Real Estate Tax	Monthly Real Estate Tax	Monthly Mortgage Payment	Monthly Bank Payment
5.	Jocelyn	$864	?	$704	?
6.	Britney	$1,380	?	$677	?
7.	Lance	$906	?	$559	?
8.	Ervan	$720	?	$442	?

9. **IN YOUR WORLD** Talk to real estate agents selling houses in your neighborhood. Find out how much you will pay in real estate tax if you buy a house. Write a report about the real estate tax for different houses in your neighborhood.

What About Homeowner's Insurance?

Every homeowner should have **homeowner's insurance.** The insurance covers the home and some of the things in it. The policy will pay for damage or loss caused by fire, smoke, theft, or some types of severe weather.

Did You Know?

Renter's insurance is very similar to homeowner's insurance. You can purchase renter's insurance to protect items you have in your rented home such as furniture and clothing.

The cost of the insurance depends on the following:

• construction of the house (brick or wood)

• location of the house (urban, suburban, or rural area)

• crime rate in the area

• location of the nearest fire hydrant

• weather patterns in the area

▶ **EXAMPLE 1**

Homeowner's insurance costs Felipe $531 a year. He needs to pay 25% of the cost as the first payment before coverage begins. How much is the first payment?

Find 25% of the cost of the insurance. First change the percent to a decimal. Then multiply.

$531 × 0.25 = $132.75

The first payment will be $132.75.

▶ **EXAMPLE 2**

Adena's homeowner's insurance costs $460 a year. Her first payment is $115. Then each payment is 10% of the balance due after the first payment. How much is each remaining payment?

STEP 1   Subtract the first payment from the cost of the insurance.

$460 − $115 = $345

STEP 2   Change the percent to a decimal. Then multiply by the balance due.

$345 × 0.10 = $34.50

Each remaining payment is $34.50.

Practice and Apply

Solve each problem. Show your work.

1. Sue buys a house. Homeowner's insurance will cost $800 a year. She needs to pay 25% of this amount as her first payment. How much is her first payment?

2. Richie buys a house. Homeowner's insurance will cost $610 a year. He needs to pay 20% of this amount as his first payment. How much is his first payment?

3. Luka buys a house. Homeowner's insurance will cost $715 a year. He needs to pay 15% of this amount as his first payment. How much is his first payment?

4. Kirsten's homeowner's insurance costs $525 a year. Her first payment is $125. Then each payment is 20% of what she owes after the first payment. How much is each remaining payment?

5. Wade's homeowner's insurance costs $650 a year. His first payment is $140. Then each payment is 20% of what he owes after the first payment. How much is each remaining payment?

6. Yvette's homeowner's insurance costs $740 a year. Her first payment is $200. Then each payment is 10% of what she owes after the first payment. How much is each remaining payment?

7. **IN YOUR WORLD** Do you think homeowner's insurance is a good idea? Why or why not?

Solve each problem. Show your work.

1. Nathan is paid 9% of the monthly rent for each apartment in the building he manages for the owner. The owner gets the rest of the rent. Nathan manages 30 apartments with a total rent of $13,500 per month. How much does the owner get each month?

2. Lavan buys a house with a down payment of $12,000. The down payment is 10% of the price of the house. How much will her mortgage be?

3. Nicole makes $34,000 a year. According to the banker's rule, can she afford to borrow $88,000 for a house? Explain.

4. **OPEN ENDED** Donato's gross monthly income is $1,500. He buys only the things he needs. How much can he afford to pay in rent?

Calculator

Al signed a one-year lease. He moved out 9 months later. The rent was $800 a month. Al used a calculator to find out how much he had to pay in rent when he moved out.

First he found the number of months left on the lease.

Press: 1 2 − 9 =   3

Then he multiplied by the monthly rent.

Press: × 8 0 0 =   2,400

Find how much rent each tenant owed when he or she moved out.

	Name	Lease	Moved Out	Monthly Rent	Rent Owed
1.	Adena	One-year	after 10 months	$900	?
2.	Juan	Two-year	after 15 months	$750	?

DECISION MAKING:
Which House Should You Purchase?

Adena found two houses she likes and can afford. Each house has 2 bedrooms.

The first house costs $76,000. It was built in 1962. There is a fireplace in the living room. There is a backyard where Adena could plant a garden. The house is not close to Adena's job or to the grocery store. However, it is a short walk to the bus stop.

The other house Adena likes costs $83,000. It was built in 1984. It does not have a fireplace or room for a garden. There are new appliances in the kitchen. Adena could walk from this house to work each day.

Adena thinks carefully about what she wants and really needs in a home. She needs to live close to work or transportation, because she does not own a car. She won't have much extra money after she pays the mortgage each month. It is important that Adena find a home that will not need major repairs. Adena would like to have a garden, and she loves fireplaces, but these are not her needs.

Adena made the following chart to compare the two homes.

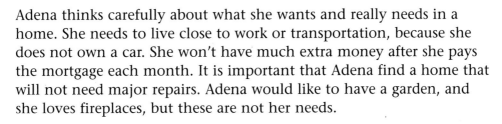

Home	Price	Number of Bedrooms	Close to Work	Year House Was Built	Extras
A	$76,000	2	no	1962	fireplace, garden
B	$83,000	2	yes	1984	new appliances

Use the chart above to answer each question.

1. Which house costs less?

2. Which house is closer to where Adena works?

3. Which house is more likely to need repairs first? Why?

You Decide

Which house should Adena choose? Be sure to take into account all of Adena's wants and needs. Explain your choice.

Summary

The rent you pay for an apartment should not be more than 25%–30% of your income.
When you look for a place to live, you need to consider the cost, the number of rooms, and the location.
A lease is a legal agreement between the renter and the landlord. Breaking a lease can cost money.
When you move out of a place you have rented, the security deposit pays for anything you damaged during the lease. The amount of your deposit not used for repairs is returned to you.
The banker's rule says that you should not borrow more than 2.5 times your annual income for your mortgage.
The down payment required on a house can be as high as 20% of the price. Closing costs are usually between 3% and 6% of the price of a house.
If you borrow money to buy a house, you pay a mortgage payment each month.
You can pay the real estate tax with your mortgage each month.
If you own a home, you should have homeowner's insurance to pay for damages or losses.

landlord

lease

real estate tax

rent

security deposit

Vocabulary Review

Complete the sentences with words from the box.

1. A legal contract between a renter and the landlord is a ____.

2. The monthly fee paid to a landlord is ____.

3. A person who owns an apartment or house and rents it to others is a ____.

4. ____ is a local tax on property that is paid by the owner.

5. Money that will be returned when the renter moves out if the place is clean and has not been damaged is the ____.

Chapter Quiz

Solve each problem. Show your work.

1. Maria's monthly income is $1,250. She can afford to spend 25% of her income for rent. How much can she pay for rent each month?

2. Look at the ad on the right. It is for an apartment on Maple Avenue. How many bedrooms does the apartment have?

3. José signed a one-year lease. He pays $560 per month in rent. José moves out 3 months early. How much money will he owe when he moves out?

4. Martin finds an apartment to rent for $420 per month. He must pay a security deposit equal to one and a half months' rent. How much is the security deposit?

5. Alexis earns $31,350 per year. According to the banker's rule, how much money can she afford to borrow for a house?

6. Larry found a house for $120,500. He pays a 20% down payment. How much does Larry pay for the down payment?

7. Aaron took out a 30-year mortgage for $70,000. His monthly mortgage payment is $466. How much will he pay over 30 years?

Maintaining Skills

Write each percent as a decimal.

1. 5.6%   2. 8%   3. 2.9%

Find each percent.

4. 6% of $140,000   5. 3.5% of $90,000   6. 5% of $120,000

Before you move into your new place, you may want to decorate. You may decide to paint the rooms. How do you know how much paint you will need?

Learning Objectives

LIFE SKILLS

- List priorities based on income and need.
- Identify ways to furnish a home on a budget.
- Find the sale price.
- Compare the efficiency ratings of appliances.
- Choose a phone plan.
- Determine how much paint is needed to paint a room.
- Determine how many tiles are needed to cover a floor.

MATH SKILLS

- Add, subtract, multiply, and divide.
- Find the perimeter and area of a rectangle.

Words to Know

priority	something that is more important than most other things
furnishings	furniture, appliances, rugs, curtains, and other items used in a home
bargain	something offered or bought at a low price; to discuss buying an item at a lower price
household necessities	items needed for the home
discount	an amount subtracted from the original price of an item
appliance	a machine designed to do tasks in the home, such as a vacuum cleaner, washing machine, or blender
energy efficiency rating	a dollar amount assigned to an appliance that shows the cost of the energy it uses on average in one year
perimeter	the distance around a figure or room
area	the number of square units that cover a surface

Project: Buying Household Items

You have $1,000 to buy household items. Choose three furniture items that you will need for one room. Decide where you will buy each item. Compare the prices at several stores. You may also look for the items on sale in the classified section of the newspaper. Create a spreadsheet on a computer to compare the price of each item at different stores and in the newspaper ads. Look for the best price for each item. Then add the total cost of the items. Did you stay within your budget of $1,000? Do you have any money left over?

10·1 ▶ What Are Your Priorities?

Kevin is furnishing his new home. He needs to decide what he can buy with a limited budget.

He thinks about the items he needs. Then, he decides if each item is a **priority**. A priority is something that is more important than most other things.

Below is a list of the things Kevin needs to buy. They are listed in order of priority. Your list of **furnishings** might look something like Kevin's.

THINGS TO BUY
1. Bed
2. Table and chairs
3. Sofa
4. Dresser
5. Television
6. Desk
7. Computer
8. Coffee table
9. CD player
10. VCR or DVD player

This list shows that having a bed is a higher priority for Kevin than having a VCR. This list will change depending on what a person needs. A computer would be a high priority for a student. A student may need a computer to do research and write papers.

Practice and Apply

Answer each question.

1. In the list above, what is Kevin's highest priority?

2. Adena wants these items: CD player, bed, television, sofa, kitchen table, lamps. List the items in the most likely order of priority. Which item is Adena's first priority?

3. **IN YOUR WORLD** Suppose your new apartment does not have a refrigerator. Why would buying a refrigerator be a high priority?

Kevin wants to pay cash for the furnishings he needs, but he doesn't have a lot of money to spend. There are many places he can find a **bargain.** Kevin might find a low-priced coffee table in:

- a used furniture store

- a garage sale

- the classified ads in the newspaper

Kevin plans to bargain with the seller when he buys items at garage sales or through classified ads. He will offer the seller less than the asking price. The seller might accept his offer or suggest a higher price that is still less than the asking price.

Here's a Tip!
In classified ads, look for *firm* or *OBO* next to the price. *Firm* means the seller will not accept a lower price. *OBO* means *or best offer.* The seller may take less than the asking price.

Practice and Apply

Solve each problem. Show your work.

1. You see a classified ad for a dresser for $100. You offer the seller $75. He rejects your offer, but offers to lower the price by $15. How much will you pay for the dresser if you accept his offer?

2. You have $117 to buy a sofa and a rug for your new apartment. At a garage sale, you see a sofa for $95. You pay $5 off the marked price of the sofa. How much money do you have left for the rug?

3. You have $192 to furnish your apartment. You buy a bed for $60, a table and chairs for $45, a chest for $32, and a microwave for $25. How much money do you have left?

4. **IN YOUR WORLD** Think of something you have that you would like to sell. Look at ads selling similar items. Write an ad to sell your item.

You will need many household items for your home. These items can cost a lot. You can save money by buying these **household necessities** on sale. The **discount** is the amount of money you save when you buy an item on sale.

▶ **EXAMPLE 1**

Eli finds sheets for $30 in a department store. Next week the sheets will be on sale for 20% off. He decides to wait for the sale to buy the sheets. How much will Eli pay for the sheets on sale?

STEP 1	Change 20% to a decimal.	20% = 0.20
STEP 2	Multiply the original price by the decimal to find the discount.	$30 × 0.20 = $6
STEP 3	Subtract the discount from the original price to find the sale price.	$30 − $6 = $24

Eli will pay $24 for the sheets on sale.

▶ **EXAMPLE 2**

Eli finds a blanket at a department store. The original price is $40. It is on sale for 10% off. He has a coupon for 15% off the sale price. How much will the blanket cost on sale with the coupon?

Here's a Tip!
Do not add the two discounts together to find the final price. The store coupon is for 15% off the sale price. So you need to find the sale price first before finding the discount from the company.

STEP 1	Change 10% to a decimal.	10% = 0.10
STEP 2	Multiply the original price by the decimal to find the discount.	$40 × 0.10 = $4
STEP 3	Subtract to find the sale price.	$40 − $4 = $36
STEP 4	Now change 15% to a decimal.	15% = 0.15
STEP 5	To find the coupon savings, multiply the sale price by the decimal.	$36 × 0.15 = $5.40
STEP 6	Subtract to find the final cost.	$36 − $5.40 = $30.60

The blanket will cost $30.60 on sale with the coupon.

Practice and Apply

Solve each problem. Show your work.

1. Cy needs to buy a package of dish towels. The original price is $15. The dish towels are now on sale for 20% off. How much will Cy pay for the dish towels on sale?

2. Brad needs to buy a phone. The original price is $36. The phone is now on sale for 30% off. How much will Brad pay for the phone on sale?

3. Darrell needs to buy bath towels. If he buys 3 towels, he will get 15% off the total cost. Each towel costs $10. How much will Darrell pay altogether if he buys 3 towels?

4. Erica finds a set of pots at a department store. The original price is $80. The pots are on sale for 15% off. She has a coupon for 10% off the sale price. How much will the pots cost on sale with the coupon?

5. Morgan finds a set of sheets at a department store. The original price is $45. They are on sale for 20% off. She has a coupon for 15% off the sale price. How much will the sheets cost?

6. Haitham bought a DVD player. The original price is $175. The DVD player was on sale for 30% off. Haitham also applied for the store's credit card that day. He got an additional 10% off the sale price of the DVD player. How much did Haitham pay for the DVD player? (Hint: First find the sale price. Then find 10% off the sale price.) Use a calculator if you like.

7. **CRITICAL THINKING** Renée saw plates at a department store. The original price was $40. The plates are on sale for 25% off. She has a coupon for an additional 25% off the sale price. The same plates cost $27 at a discount store. How much are the plates at the department store? Where should Renée buy the plates?

Before you buy a large **appliance**, such as a refrigerator, you should find out about its **energy efficiency rating**. Choosing an appliance with a good energy rating can save you a lot of money on utility costs. It is also better for the environment because the appliance uses less energy.

Did You Know?
The cost of electricity is given in kilowatt-hours. A kilowatt-hour is the amount of power supplied by one kilowatt for one hour.

The refrigerator energy guide below shows a rating of $91 at 5 cents per kilowatt-hour. This is the average cost of the electricity needed to operate this refrigerator for one year. The Yearly Cost chart in the energy guide below shows the yearly cost for different electricity rates.

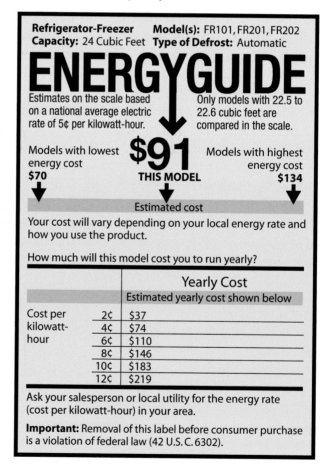

Refrigerator-Freezer **Model(s):** FR101, FR201, FR202
Capacity: 24 Cubic Feet **Type of Defrost:** Automatic

ENERGYGUIDE

Estimates on the scale based on a national average electric rate of 5¢ per kilowatt-hour.

Only models with 22.5 to 22.6 cubic feet are compared in the scale.

Models with lowest energy cost
$70

$91
THIS MODEL

Models with highest energy cost
$134

Estimated cost

Your cost will vary depending on your local energy rate and how you use the product.

How much will this model cost you to run yearly?

		Yearly Cost
		Estimated yearly cost shown below
Cost per kilowatt-hour	2¢	$37
	4¢	$74
	6¢	$110
	8¢	$146
	10¢	$183
	12¢	$219

Ask your salesperson or local utility for the energy rate (cost per kilowatt-hour) in your area.

Important: Removal of this label before consumer purchase is a violation of federal law (42 U.S.C. 6302).

Practice and Apply

Use the energy guide on page 204 to solve each problem.

1. What is the energy efficiency rating for this refrigerator?

2. How much would it cost a year to operate a similar model with the lowest energy cost?

3. How much would it cost a year to operate a similar model with the highest energy cost?

4. How much would it cost a year to operate this refrigerator if you lived in an area that charges 4¢ per kilowatt-hour of electricity?

5. Suppose you kept this refrigerator for 12 years. What would your cost be to operate it over that length of time if the electricity cost per kilowatt-hour in your area is 10¢? (Hint: Multiply the yearly cost by 12 years.)

6. Suppose you kept this refrigerator for 10 years. What would your cost be to operate it over that length of time if the electricity cost per kilowatt-hour in your area is 6¢?

7. **CRITICAL THINKING** Cary and Kim both bought new refrigerators this year. They live in the same town. Why might Cary pay more in electricity costs for the new appliance than what Kim pays?

Maintaining Skills

Change each decimal to a percent.

1. 0.325   **2.** 0.06   **3.** 0.67   **4.** 1.25

10·5 Which Plan Is Best for You?

When you are on your own, you will have to pay for telephone service. You may be able to choose which long distance service is best for you. You will need to compare the service plans. The plans will charge different rates for calls you make on weekdays, and calls you make on evenings and weekends. The chart below shows two different plans.

Phone Plans		
Charges	Plan A	Plan B
Monthly fee	None	$14.95
Price per minute weekdays	$0.15	Free first 100 minutes Then $0.04
Price per minute evenings/weekends	$0.10	Free first 200 minutes Then $0.03

To decide between the plans, you need to know when you make calls and about how long the calls are.

▶ **EXAMPLE**

Eric thinks that each month he will talk on the phone for about 35 minutes on weekdays, and about 75 minutes on evenings and weekends. Which plan is better for Eric?

STEP 1 Find the cost of each type of call with Plan A.
Weekdays   Evenings/Weekends
35 × $0.15 = $5.25   75 × $0.10 = $7.50

STEP 2 Add to find the total cost of Plan A.
$5.25 + $7.50 = $12.75

STEP 3 Find the cost of each type of call with Plan B.
Weekdays   Evenings/Weekends
35 < 100 ⟶ Free   75 < 200 ⟶ Free

STEP 4 Add the monthly fee to find the cost of Plan B.
$0 + $14.95 = $14.95

STEP 5 Compare the costs of the plans.
$12.75 < $14.95

Plan A would cost less. Plan A is better for Eric.

Practice and Apply

Find the cost of each plan. Use the chart on page 206.

	Name	Weekday Minutes Used	Evening/Weekend Minutes Used	Cost of Plan A	Cost of Plan B
1.	Juan	40	80	?	?
2.	Tak	40	85	?	?
3.	Aaron	50	90	?	?
4.	Mari	80	150	?	?
5.	Nicole	90	210	?	?

Use the chart on the right to answer each question.

6. How much does the Basic plan cost?

7. Gavin does not watch TV very often. His favorite weekly shows are on the basic channels. Which plan is best for him?

Cable Plans	
Basic	$9.99
Extended* (2 movie channels)	$29.99
Premium* (all movie channels)	$42.99
*Basic cable included in these plans	

8. Ross likes to watch movies. He has $30 in his budget for cable service. Which plan should he pick?

9. Taryn needs to watch a lot of movies for a film course he is taking. He has $45 in his budget for cable service. Which plan should he pick?

10. **WRITE ABOUT IT** Mari lives on her own. She is a student at a community college. She has a part-time job. Which cable television plan should she choose? Explain.

Suppose you want to put a fence around your garden. You will need to find the distance around the garden or the **perimeter.** You can use a formula.

Perimeter = (2 × length) + (2 × width)   width [ ]

length

▶ **EXAMPLE 1**

Marcy wants to build a pen for her dog. The pen will be a rectangle 20 feet long and 12 feet wide. How many feet of fencing does Marcy need for the pen?

STEP 1 Write the formula.   Perimeter = (2 × length) + (2 × width)

STEP 2 Write 20 for length and 12 for width.   Perimeter = (2 × 20) + (2 × 12)

STEP 3 Multiply. Then add.   Perimeter = 40 + 24 = 64

Marcy needs 64 feet of fencing for the pen.

You may want to paint the wall of a room. You will need to find the number of square units needed to cover the wall, or the **area.** You can use a formula.

Area = length × width   width [ ]

length

▶ **EXAMPLE 2**

Here's a Tip!
Area is measured in square units.

Glen wants to paint a wall that is 12 feet long and 8 feet high. The wall does not have any windows or doors. How much area will the paint have to cover? (Hint: Use the height of the wall as the width.)

STEP 1 Write the formula.   Area = length × width

STEP 2 Write 12 for length and 8 for width.   Area = 12 × 8

STEP 3 Multiply.   Area = 96

The paint will have to cover 96 square feet.

Practice and Apply

Find the perimeter and area of each rectangle.
(Hint: Opposite sides of a rectangle are equal.)

	Length	Width	Perimeter	Area
1.	10 inches	5 inches	?	?
2.	4 feet	8 feet	?	?
3.	15 meters	2 meters	?	?
4.	9 feet	8 feet	?	?
5.	20 inches	4 inches	?	?

Everyday Problem Solving

Solve each problem. Show your work.

1. Brian needs to build a fence around the garden. The plan for Brian's garden is on the right. How many feet of fencing does he need? (Hint: Subtract the length of the gate from the perimeter.)

3 ft
gate

10 ft

15 ft

2. What is the area of Brian's garden? (Hint: The gate does not affect the area.)

3. Kendra puts up 50 feet of fencing to separate her yard from her neighbor's yard. Fencing costs $13 a foot. How much will the fencing cost?

4. Shawn needs 40 feet of fencing to build a pen for his dog. Fencing costs $11 a foot. The gate costs $15. How much will the fence and the gate for the pen cost altogether?

Mikel wants to paint her living room and bedroom. She wants to find out how much paint she will need.

Paint is sold by the quart and by the gallon. First Mikel needs to know the number of square feet to be painted. She must also find out the number of square feet the paint will cover. Then she can find out how much paint she should buy.

▶ **EXAMPLE**

The area Mikel will paint is 650 square feet. One gallon of paint covers 400 square feet. One quart of paint covers 100 square feet. How much paint does she need?

Here's a Tip!
4 quarts = 1 gallon

STEP 1 Divide the total area of the rooms by the area one gallon covers.

$$\begin{array}{r} 1 \text{ R250} \\ 400\overline{)650} \\ -400 \\ \hline 250 \end{array}$$

Mikel needs 1 gallon of paint. But this is not enough. She still needs to paint 250 square feet more.

STEP 2 Divide 250 by the area one quart covers.

$$\begin{array}{r} 2 \text{ R50} \\ 100\overline{)250} \\ -200 \\ \hline 50 \end{array}$$

Mikel needs 3 quarts of paint because 2 quarts will not be enough. Altogether Mikel needs 1 gallon and 3 quarts of paint.

It may cost less for Mikel to buy 2 gallons of paint instead of 1 gallon and 3 quarts. Suppose 1 gallon of paint costs $12 and 1 quart of paint costs $5.

Cost of 2 gallons = 2 × $12
= $24

Cost of 1 gallon and 3 quarts = $12 + 3 × $5
= $12 + $15
= $27

$24 < $27. It costs less to buy 2 gallons of paint.

Practice and Apply

Solve each problem. Show your work.

1. Sally wants to paint her kitchen. The area to be painted is 260 square feet. One quart of paint covers 90 square feet. How many quarts of paint does Sally need?

2. Darrel wants to paint his living room. The area to be painted is 580 square feet. One gallon of paint covers 400 square feet. One quart of paint covers 100 square feet. How much paint does he need?

3. Aaron wants to paint his bedroom and hallway. The area to be painted is 750 square feet. One gallon of paint covers 400 square feet. How many gallons of paint does he need to buy?

4. Juan needs 3 quarts of paint to paint his bedroom. A quart of paint costs $3.89. How much will it cost Juan to buy the paint for his bedroom?

5. Ebony needs 1 gallon and 2 quarts of paint to paint her bedroom. A gallon of paint costs $14.08. A quart of paint costs $5.89. How much will it cost Ebony to buy the paint for her bedroom?

6. Sun Li needs 2 gallons and 3 quarts of paint to paint her living room. A gallon of paint costs $12.97. A quart of paint costs $4.85. How much will it cost Sun Li to buy the paint for her living room? Use a calculator if you like.

7. **CRITICAL THINKING** Luis wants to paint his bedroom. He needs to estimate the number of square feet he will paint before he begins. Each of the four walls is 12 feet long and 8 feet high. Estimate how many square feet he will paint. (Hint: Find the area of each wall. Then, add the areas to find the total.)

10·8 How Many Tiles Do You Need to Cover the Floor?

Suppose you want to put tile down on a kitchen or bathroom floor. First you need to find out how many tiles you will need to cover the floor. Then if you know the cost of 1 tile, you can find the total cost of the tiles you need.

▶ **EXAMPLE 1**

Ron wants to put tile down on his kitchen floor. The floor is 12 feet long and 8 feet wide. Each tile covers 1 square foot. The tiles cost $1.25 each. What is the total cost of the tiles for Ron's kitchen floor?

STEP 1 Find the area of the floor. **Multiply the length by the width.**

12 × 8 = 96 square feet

STEP 2 Determine the number of tiles needed. **Each tile is 1 square foot.**

96 square feet ⟶ 96 tiles

STEP 3 Multiply the number of tiles by the cost of each tile.

96 × $1.25 = $120.00

The total cost of the tiles for Ron's kitchen floor is $120.

Sometimes tiles are sold by the box. If you know how many tiles are in a box and how many tiles you need, you will know how many boxes of tile to buy.

▶ **EXAMPLE 2**

Krista wants to put tile down on her bathroom floor. She needs 36 tiles. There are 24 tiles in a box. How many boxes of tile should Krista buy?

Divide the number of tiles needed by the number of tiles in a box.

36 ÷ 24 = 1.5

Krista needs 1.5 boxes of tiles. She cannot buy half a box. Krista should buy 2 boxes of tiles.

Practice and Apply

Solve each problem. Show your work.

1. Jen wants to put tile down on her kitchen floor. The floor is 16 feet long and 10 feet wide. Each tile covers 1 square foot. The tiles cost $1.95 each. What is the total cost of the tiles for Jen's kitchen floor?

2. Raul wants to put tile down on his bathroom floor. The floor is 8 feet long and 5 feet wide. Each tile covers 1 square foot. The tiles cost $1.20 each. What is the total cost of the tiles for Raul's bathroom floor?

3. Brian wants to put tile down on his kitchen floor. He needs 84 tiles. There are 24 tiles in a box. How many boxes of tile should Brian buy?

4. Janet wants to put tile down on her bathroom floor. She needs 45 tiles. There are 18 tiles in a box. How many boxes of tile should Janet buy?

5. Mari wants to put tile down on her kitchen floor. The floor is 12 feet long and 10 feet wide. Each tile covers 1 square foot. There are 20 tiles in a box. Each box costs $15.25. How much will the tiles for Mari's kitchen floor cost?

6. Susan wants to put tile down on her bathroom floor. The floor is 8 feet long and 6 feet wide. Each tile covers 1 square foot. The tiles are sold in boxes. There are 18 tiles in a box. Each box costs $21.50. How much will the tiles for Susan's kitchen floor cost? Use a calculator if you like.

7. **CRITICAL THINKING** Martin wants to put tile down on his kitchen floor. The floor is 9 feet long and 8 feet wide. Each tile is 6 inches by 6 inches. How many tiles does he need to cover the floor? (Hint: 4 tiles cover 1 square foot. Multiply the area in square feet by 4.)

10·9 ▶ Problem Solving

Solve each problem. Show your work.

1. Gena finds a kitchen table and 4 chairs at a yard sale. The seller is asking for $50 for the table and $10 for each chair. She offers $75 for the whole set. The seller agrees. How much does Gena save?

2. Tami needs to buy a toaster. The original price is $28. The toaster is now on sale for 15% off. How much will Tami pay for the toaster on sale?

3. What is the area of a garden that is 16 feet long and 11 feet wide?

4. Howard wants to put tile down on his kitchen floor. The floor is 12 feet long and 9 feet wide. Each tile covers 1 square foot. The tiles cost $1.50 each. How much will the tiles for Howard's kitchen floor cost?

5. **OPEN ENDED** The area of a floor is 100 square feet. What could be the length and width of the room? (Hint: The product of the length and width must be 100.)

Calculator

On some calculators, order of operations is important.

Suppose one shirt costs $24. The second shirt is half price. What is the total cost of two shirts?

First divide 24 by 2 to find the cost of the second shirt.

Then add the cost of the first shirt to find the total cost.

Press: 2 4 ÷ 2 + 2 4 = ⟦ 36 ⟧

Find the total cost for two of each item. You pay full price for one item and half price for the second item.

1. A shirt is $28.   2. A T-shirt is $16.   3. A pair of pants is $32.

ON-THE-JOB MATH:
Professional Painter

Jay is a house painter. Jay writes an estimate for each customer before he starts a job. He estimates the cost of the paint and the cost of labor.

To estimate the cost of the paint, Jay needs to know how many square feet are to be painted. He needs to know how many square feet a gallon of paint covers. He needs to know how much a gallon of paint costs.

To estimate the cost of labor, Jay needs to determine about how many hours it will take to paint the room. If the room has a lot of windows and doors, the job will be more difficult and it may take longer.

Solve each problem. Show your work.

1. Mr. Silva wants Jay to paint his kitchen. Jay began to estimate the cost of the job. Complete the estimate on the right. Find the cost of the paint and the cost of the labor. Then find the total.

2. Mr. Sanchez wants Jay to paint his living room. He will need 3 gallons of paint. The paint costs $16 per gallon. The job will take about 15 hours. Jay charges $20 an hour. Write an estimate for the job.

Estimate

Job: Mr. Silva's kitchen

Paint
    2 gallons × $18/gallon = _____

Labor
    19 hours × $20/hour   = _____

                            Total

You Decide
Business is slow in January, so, Jay offers a discount on his painting. He will give you 30% off the cost of paint or 5% off the cost of labor. The customer can choose which discount to take. If you were Jay's customer, which discount would you choose? Why? (Hint: Look at the job estimates for exercises 1 and 2.)

Summary

Decide on your priorities before you purchase household items.
You can look in yard sales, used furniture stores, and the classified ads in the newspaper for household items.
To find the sale price, first find the discount. Then subtract the discount from the original price.
When you buy an appliance, choose an appliance that has a good energy efficiency rating.
To choose a phone plan, you need to know about how many minutes you will be on the phone each month. You also need to know when you will make the calls.
To find the perimeter of a rectangle, use the formula Perimeter = (2 × length) + (2 × width).
To find the area of a rectangle, use the formula Area = length × width.
You can find how much paint you need to paint a room. Divide the area to be painted by the area a quart or a gallon of paint will cover.
You can find how many tiles you need to cover a floor. You need to find the area of the floor.

appliance
bargain
furnishings
priority

Vocabulary Review

Complete the sentences with words from the box.

1. A _____ is something that is more important than most other things.

2. Furniture, appliances, rugs, curtains, and other items used in a home are _____.

3. A machine designed to do tasks in the home, such as a vacuum cleaner, washing machine, or blender, is an _____.

4. Something offered or bought at a low price is a _____.

Chapter Quiz

Solve each problem. Show your work. Round to the nearest
cent if necessary.

1. You go to a yard sale. There is a table for $80, dishes for $18, and drinking glasses for $12. The seller offers to give you the drinking glasses for free if you buy the table and dishes. You agree. How much money will you spend?

2. Corina needs new bath towels. The towels she wants are on sale for 30% off. The original price of each towel is $12.50. How much will she pay for 3 towels on sale?

3. Abe sees a blanket and pillow set for $99. The set is on sale for 15% off. He has a coupon for an additional 10% off the sale price. How much will Abe spend for the blanket and pillow set?

4. Darcy signs up for a new Internet plan. She pays $4.99 for 8 hours of Internet service a month. She pays $1.49 for each additional hour over 8 hours. How much will her bill be if she uses the Internet for 12 hours this month?

5. Rick wants to put tile down on his kitchen floor. The floor is 15 feet long and 18 feet wide. The tiles he chose are 1 square foot each. How many tiles does Rick need?

Maintaining Skills

Multiply.

1. $18 × 19   2. $1.15 × 142   3. $22 × 14   4. 45 × $0.95

Find each percent.

5. 15% of $112   6. 20% of $158

7. 10% of $99   8. 25% of $64

Unit 5 Review

Write the letter of the correct answer. Use the chart to answer questions 1 and 2.

Monthly Expenses

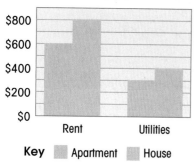

Key ▢ Apartment ▢ House

1. How much does it cost to rent the apartment for a year?
   A. $600
   B. $7,200
   C. $9,600
   D. $10,000

2. How much does it cost for utilities on the house for a year?
   A. $200
   B. $400
   C. $2,400
   D. $4,800

3. Darlene needs a 20% down payment for the house she is buying. The house costs $66,000. How much money does Darlene need for the down payment?
   A. $1,320
   B. $52,800
   C. $13,200
   D. $79,200

4. A phone company charges $0.15 per minute for weekdays and $0.10 per minute for evenings/weekends. Each month Ray talks on the phone for about 40 minutes weekdays and 90 minutes evenings/weekends. About how much will these calls cost altogether?
   A. $15.00
   B. $19.50
   C. $25.00
   D. None of the above

5. Liuda wants to put tile down on her kitchen floor. The floor is 14 feet long and 12 feet wide. What is the area of the floor?
   A. 52 square feet
   B. 168 square feet
   C. 196 square feet
   D. 200 square feet

Challenge

Luz wants to buy a house that costs $145,000. She has $55,000 for the down payment. What is the amount of Luz's mortgage? Luz gets a mortgage for 30 years at 8% interest. What is Luz's monthly mortgage payment? Use the chart on page 186.

Chapter 11 **Eating for Good Health**

Chapter 12 **Choosing and Buying Groceries**

You need milk and other foods to stay healthy. What you need and how much you need depends on your gender and age.

The circle graph shows the different daily food requirements for teenage girls. Use the graph to answer each question.

Daily Serving Requirements

18 yr. old Girl

Number of Servings

Key
Bread/Cereal
Vegetables
Fruit
Milk
Meat/Poultry/Fish

1. Which food group should teenage girls eat the most servings from daily?

2. What is the total number of daily servings required for a teenage girl?

3. How many servings of vegetables should a teenage girl eat daily?

4. Which food group should a teenage girl eat the least number of servings from daily?

It is important to buy the foods you need for a healthy diet. What are some foods and vegetables you like to eat?

Learning Objectives

LIFE SKILLS

- Learn about the food groups and plan a balanced diet.

- Find the amount of protein a person needs.

- Count calories and identify activities that burn calories.

- Identify the health problems caused by eating too much fat, sugar, or salt.

MATH SKILLS

- Solve a proportion.

- Add, subtract, multiply, and divide decimals.

- Find the percent of a number.

Words to Know

food groups	the food categories used to plan a balanced diet
diet	everything that you eat and drink regularly
nutritious	having what is needed for growth and good health
serving	a portion of food or drink usually eaten at one time
vegetarian	a person who does not eat meat
protein	a substance that is found in the cells of animals and plants; needed for the growth and repair of body cells
calorie	a unit of energy that comes from food
ratio	a comparison of one amount with another amount
proportion	a statement that two ratios are equal
cross products	the results of cross multiplying
saturated fat	a fat such as butterfat or chicken fat, which comes from an animal and is solid at room temperature
unsaturated fat	an oil such as olive oil, corn oil, or safflower oil that is liquid at room temperature

Project: Creating a Healthy Menu

Create a healthy menu for two meals and three snacks to be served for a one-day school event. Visit the United States Department of Agriculture Web site. www.usda.gov

Compare your menu with those of your classmates. Then answer the following questions.
- Which menu sounds the most appealing?
- Which menu is the least expensive?
- Which menu is the most nutritious?

What Is a Balanced Diet?

The Food Guide Pyramid below shows the different **food groups**. It tells you how much food from each group you need daily for a balanced **diet**. A balanced diet is made up of **nutritious** foods. Nutritious foods are good for your health.

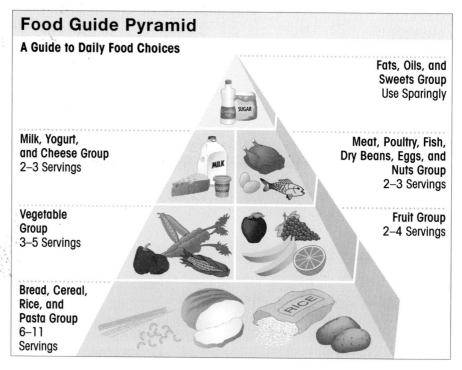

Food Guide Pyramid

A Guide to Daily Food Choices

Fats, Oils, and Sweets Group
Use Sparingly

Milk, Yogurt, and Cheese Group
2–3 Servings

Meat, Poultry, Fish, Dry Beans, Eggs, and Nuts Group
2–3 Servings

Vegetable Group
3–5 Servings

Fruit Group
2–4 Servings

Bread, Cereal, Rice, and Pasta Group
6–11 Servings

Wordwise

A *balanced diet* is one that contains foods from the different food groups.

You should eat at least the minimum number of servings from each food group daily. The chart on page 223 shows the size of one **serving** for some foods in the five major food groups.

Many prepared foods have added fat, oil, or sugars. These foods include soft drinks, sweet desserts, and fried foods. Some foods, such as meat and milk products, are naturally higher in fat than vegetables and grains. Fats, oils, and sweets should be eaten only in small amounts.

Size of One Serving				
Milk Group	Meat/Poultry/Fish Group	Vegetable Group	Fruit Group	Bread/Cereal Group
1 cup of milk	2–3 ounces of cooked lean meat, poultry, or fish	1 cup of raw leafy vegetables	1 medium apple, banana, orange, or pear	1 slice of bread
1 cup of yogurt	$1-1\frac{1}{2}$ cups of cooked dry beans	$\frac{1}{2}$ cup of other vegetables — cooked or raw	$\frac{1}{2}$ cup of chopped, cooked, or canned fruit	1 cup of ready-to-eat cereal
1.5 ounces of natural cheese	2–3 eggs	$\frac{3}{4}$ cup of vegetable juice	$\frac{3}{4}$ cup of fruit juice	$\frac{1}{2}$ cup of rice
2 ounces of processed cheese	4–6 tablespoons of peanut butter			$\frac{1}{2}$ cup of pasta

Practice and Apply

Use the chart on page 222 or the chart above to solve each problem.

1. Your greatest number of food servings should come from which food group?

2. How many servings from the vegetable group should you eat every day?

3. What is the size of 1 serving of milk?

4. What is the size of 1 serving of cooked vegetables?

5. Brian made a peanut butter sandwich. He spread 4 tablespoons of peanut butter on 2 slices of whole wheat bread. How many servings from each food group were in the sandwich?

6. For breakfast, Lorraine had 2 eggs, 1 slice of bread, $\frac{3}{4}$ cup of orange juice, and a banana. What is the minimum number of servings she still needs from each food group for the day? (Hint: 2 eggs is 1 serving for the meat group.)

7. **IN YOUR WORLD** Write down everything you eat today. Name the food group to which each food belongs. Did you eat a balanced diet? Explain.

What Can I Do If I Do Not Eat Some Foods?

Suppose you are a **vegetarian.** A vegetarian does not eat meat. Meat is very high in **protein.** Your body needs a certain amount of protein each day. The chart below will help you find the amount of protein you need each day.

Amount of Protein a Person Needs Each Day			
Age in Years	Weight	Grams of Protein per Pound	Daily Protein Need in Grams
11–14	?	× 0.45	?
15–18	?	× 0.39	?
19+	?	× 0.36	?

EXAMPLE

Hank is 17 years old. He weighs 150 pounds. How much protein does Hank need every day?

STEP 1 Look at the chart above. Find how much protein a 17-year-old needs.   0.39 gram per pound

STEP 2 Multiply the person's weight by the protein need. $150 \times 0.39 = 58.5$

Hank needs 58.5 grams of protein each day.

You can get protein from foods other than meat. See the chart below.

Food	Amount of Protein
1 cup soy milk	6.74 grams
1 cup oatmeal	6.08 grams
3 ounces canned tuna	20.08 grams
1 cup kidney beans	15.35 grams
1 cup instant white rice	3.4 grams

Practice and Apply

Use the charts on page 224 to solve each problem. Show your work.

1. How much protein does a 19-year-old who weighs 128 pounds need each day?

2. How much protein does a 12-year-old who weighs 90 pounds need each day?

3. Marcus is 22 years old. He weighs 165 pounds. He has consumed 45 grams of protein today. How many more grams of protein must he consume today to meet his daily need?

4. Dorie is 18 years old. She weighs 140 pounds. Today she consumed 12 grams of protein at breakfast. She consumed 28 grams of protein at lunch. How many more grams of protein must she consume to meet her daily need?

5. Aaron is 16 years old. He weighs 135 pounds. He had 1 cup of soy milk with 1 cup of oatmeal for breakfast. He had 3 ounces of tuna fish in a sandwich for lunch. How many grams of protein did he consume? How many more grams of protein does he need for the day? (Hint: Use the bottom chart on page 224 to find the amount of protein in the food Aaron ate.)

6. **IN YOUR WORLD** Calculate the amount of protein you need each day. Make a list of the foods you could eat that contain protein. Explain how you could meet your daily protein need.

What Are Calories?

The foods you eat give your body energy. The energy in food is measured in a unit called a **calorie.** You can count calories to be sure that your body gets the energy it needs. The charts below list the number of calories in some foods.

Number of Calories in Food		
Food	Serving Size	Calories
American cheese	1 ounce	82
Apple	1 small	81
Bacon	3 slices	109
Baked potato	1 medium	145
Butter	1 tablespoon	102
Chicken	4 ounces	161
Chocolate cake	1 slice	235
Corn on the cob	1 medium	59
Cucumber	1 large	39
Grapefruit	$\frac{1}{2}$ medium	37
Lean beef	3 ounces	166

Number of Calories in Food		
Food	Serving Size	Calories
Lettuce	10 leaves	20
Lowfat fruit yogurt	8 ounces	232
Milk	1 cup	102
Oatmeal cookies	1 cookie	113
Scrambled egg	1 medium	101
Spaghetti	1 cup	197
String beans	1 cup	44
Tomato	1 medium	26
White rice	1 cup	162
White roll	1 large	167
Whole wheat bread	1 slice	65

▶ **EXAMPLE**

Did You Know?
Each day you need calories from each of the food groups.

Suppose you ate 1 scrambled egg, 1 slice of whole wheat bread, 1 cup of milk, and $\frac{1}{2}$ grapefruit. How many calories did you consume for breakfast?

Add to find the total number of calories in the food.

101 + 65 + 102 + 37 = 305

You consumed 305 calories for breakfast.

Practice and Apply

Use the charts on page 226 to solve each problem.
Show your work.

1. How many calories are in 1 cup of string beans?

2. How many calories are in a baked potato with
   1 tablespoon of butter?

3. How many more calories are in a slice of chocolate cake
   than in an oatmeal cookie?

4. How many calories are in 3 ounces of American cheese?

5. Margot made a sandwich on 2 slices of whole wheat
   bread. How many more calories would she consume if
   she made her sandwich on a white roll?

6. Darren went out for breakfast. He ate 3 slices of bacon,
   2 scrambled eggs, 2 slices of whole wheat bread, and
   1 cup of milk. How many calories did he consume?

7. Ruth had a salad with 10 leaves of lettuce, 1 medium
   tomato, and $\frac{1}{2}$ cucumber. She also had 1 cup of string
   beans, 1 baked potato with 1 tablespoon of butter, and
   6 ounces of lean beef. Then, she had 3 oatmeal cookies
   and 1 cup of milk. How many calories did she consume?
   Use a calculator if you like.

8. **CRITICAL THINKING** A teenage girl needs at least
   419 calories from vegetables each day. Using the foods
   in the chart, what vegetables could she eat to meet
   her needs?

Maintaining Skills

Multiply.

1. 12×8 2. 15×9 3. 20×7 4. 35×6 5. 78×3

6. 25×12 7. 30×8 8. 18×5 9. 28×6 10. 32×4

You can use a **ratio** to compare two amounts. For example, there are 232 calories in 8 ounces of lowfat fruit yogurt. You can write this as a ratio.

$$\frac{232 \text{ calories}}{8 \text{ ounces}}$$

A **proportion** shows two equal ratios.

$$\frac{232 \text{ calories}}{8 \text{ ounces}} = \frac{29 \text{ calories}}{1 \text{ ounce}}$$

The **cross products** in a proportion are equal.

$$\frac{232}{8} = \frac{29}{1}$$

$$232 \cdot 1 = 8 \cdot 29$$

$$232 = 232$$

You can use a proportion to solve a word problem.

▶ **EXAMPLE**

Here's a Tip!
Arrows can help you identify the cross products in a proportion.

$$\frac{176}{4} \diagup\hspace{-0.9em}\diagdown \frac{x}{6}$$

The cross products are 176 • 6 and 4 • x.

There are 176 calories in 4 ounces of lean pork. How many calories are there in 6 ounces of lean pork?

STEP 1 Write a proportion. Use the variable x for the unknown number of calories.

$$\text{calories} \rightarrow \frac{176}{4} = \frac{x}{6} \leftarrow \text{calories}$$
$$\text{ounces} \rightarrow \quad\quad\quad \leftarrow \text{ounces}$$

STEP 2 Write the cross products equal to each other.

$$176 \cdot 6 = 4 \cdot x$$

STEP 3 Multiply.

$$1{,}056 = 4x$$

STEP 4 Divide to find x.

$$1{,}056 \div 4 = 4x \div 4$$
$$264 = x$$

There are 264 calories in 6 ounces of lean pork.

Skills Practice

Solve each proportion.

1. $\frac{3}{4} = \frac{x}{8}$

2. $\frac{5}{x} = \frac{20}{12}$

3. $\frac{x}{9} = \frac{22}{18}$

4. $\frac{x}{5} = \frac{24}{6}$

5. $\frac{7}{11} = \frac{x}{33}$

6. $\frac{15}{x} = \frac{30}{60}$

7. $\frac{10}{x} = \frac{35}{21}$

8. $\frac{6}{13} = \frac{24}{x}$

9. $\frac{5}{x} = \frac{15}{36}$

10. $\frac{3}{5} = \frac{x}{22}$

11. $\frac{14}{15} = \frac{7}{x}$

12. $\frac{x}{20} = \frac{845}{65}$

Everyday Problem Solving

Miss Power's class has planned the cafeteria's menu for Monday. They created a menu with the serving size and number of calories in each item. Use the menu to solve each problem. Show your work.

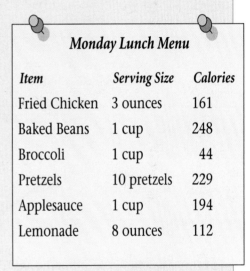

Monday Lunch Menu

Item	Serving Size	Calories
Fried Chicken	3 ounces	161
Baked Beans	1 cup	248
Broccoli	1 cup	44
Pretzels	10 pretzels	229
Applesauce	1 cup	194
Lemonade	8 ounces	112

1. How many calories would you consume if you ate one serving of each menu item?

2. How many calories are in a 6-ounce glass of lemonade? (Hint: The serving size on the menu is 8 ounces.)

3. How many calories are in 8 pretzels?

4. How many calories are in 6 ounces of chicken?

5. How many calories are in $\frac{1}{2}$ cup of applesauce? (Hint: Write 0.5 for $\frac{1}{2}$.)

6. One quart equals 32 ounces. How many calories are in a quart of lemonade?

Wordwise

The phrase *burn calories* means to use up calories.

The number of calories you need during the day depends on how active you are. If you consume more calories than you need, you gain weight. But you can exercise to burn extra calories. The charts below show an average number of calories a person can burn during certain activities. The actual number of calories you burn will depend on several things, including your weight.

Activity	Calories Burned per Hour
Watching TV	72
Karate	441
Dancing	342
Walking	297

Activity	Calories Burned per Hour
Hiking	414
Inline skating	477
Bowling	198
Volleyball	207

To keep your weight the same, you must consume as many calories as you burn in a day.

▶ **EXAMPLE**

How many more calories can a person burn bowling for 1.5 hours than watching television?

STEP 1 Multiply to find the number of calories burned bowling.

$198 \times 1.5 = 297$

STEP 2 Multiply to find the number of calories burned watching television.

$72 \times 1.5 = 108$

STEP 3 Subtract to find the difference in the number of calories burned.

$297 - 108 = 189$

A person can burn 189 more calories bowling for 1.5 hours than watching television.

Practice and Apply

Use the charts on page 230 to find the calories you can burn with each activity. (Hint: $1\frac{1}{2}$ hours is the same as 1.5 hours.)

	Activity	Total time doing the activity	Calories burned
1.	Volleyball	1 hour	?
2.	Karate	1 hour	?
3.	Walking	2 hours	?
4.	Inline skating	2 hours	?
5.	Bowling	$1\frac{1}{2}$ hours	?
6.	Dancing	$1\frac{1}{2}$ hours	?
7.	Watching TV	$\frac{1}{2}$ hour	?
8.	Hiking	$\frac{1}{2}$ hour	?

9. How many more calories can you burn by skating for 1 hour than by walking for 1 hour?

10. How many more calories can you burn by dancing for 2 hours than by bowling for 2 hours?

11. **IN YOUR WORLD** Describe a meal you often eat for lunch. Find the number of calories in that meal. Look at the chart on page 230. Choose three activities from the chart. How long would you have to do each activity to burn the calories in that meal?

What Is Wrong With Fat, Salt, and Sugar?

Too much fat, salt, or sugar is not good for you.

There are two kinds of fat. **Saturated fat** comes from animal fat. **Unsaturated fat** comes from vegetable oils. Unsaturated fats are better for you. Eating too much saturated fat may lead to heart disease. Less than 30% of your daily calories should be from fat. There are 9 calories in a gram of fat.

Did You Know?
2,400 milligrams of salt is about one teaspoon of salt.

Eating too much salt can increase your blood pressure. You should eat less than 2,400 milligrams of salt a day. Salt has no calories.

Eating too much sugar can cause weight gain and dental problems. One gram of sugar has 4 calories.

▶ **EXAMPLE 1**

Here's a Tip!
Remember that fat has 9 calories per gram.

Kofi ate a serving of corn chips. There are 10 grams of fat in the corn chips. How many of the calories in the corn chips are from fat?

Multiply the number of grams by the number of calories in a gram. $10 \times 9 = 90$

There are 90 calories from fat in the corn chips.

▶ **EXAMPLE 2**

Kofi is 16 years old. He eats about 2,800 calories daily. One day his meals included 420 calories from fat. What percent of the calories were from fat?

STEP 1 Divide to find the percent of calories from fat. $\frac{420}{2,800} = 0.15$

STEP 2 Change the decimal to a percent. $0.15 = 15\%$

15% of the calories were from fat.

Practice and Apply

Solve each problem. Show your work.

1. Which would you choose for good nutrition, a serving of frozen peas with butter or fresh peas with unsaturated margarine? Explain.

2. Which would you choose for good nutrition, a plain cereal with fresh fruit or cereal coated with sugar? Explain.

3. Which contains the greatest number of calories per gram: fat, salt, or sugar?

4. How much salt should a person eat in a day?

5. One gram of sugar has 4 calories. A can of chili with beans has 10 grams of sugar. There are 540 calories in the can. What percent of the calories in the can is from sugar? Round to the nearest whole percent.

6. One gram of fat has 9 calories. A one-ounce slice of American cheese has 9 grams of fat. The slice has a total of 110 calories. What percent of the calories in the cheese is from fat? Round to the nearest whole percent.

7. Sean ate a piece of pie that had 19 grams of fat. How many of the calories in the serving he ate were from fat?

8. **CRITICAL THINKING** Julie is 17 years old. She needs about 2,200 calories a day. Her meals one day included 800 calories from fat. Did Julie eat too much fat that day? Explain. (Hint: Less than 30% of your daily calories should be from fat.)

Solve each problem. Show your work.

1. There are 232 calories in 8 ounces of yogurt. How many calories are in 5 ounces of yogurt?

2. Students multiplied their weights by 0.39 to find how many grams of protein they need daily. Bill weighs 200 pounds. He wrote 0.78 gram. Caroline weighs 125 pounds. She wrote 48.75 grams. Who made a mistake? Write the correct answer. Explain.

3. **OPEN ENDED** You want to eat a balanced dinner that has no more than 800 calories. The menu includes a serving of turkey with 221 calories, a serving of meatloaf with 332 calories, a serving of chicken with 284 calories, breadsticks that have 46 calories each, and cookies that have 98 calories each. What would you eat? Explain.

Calculator

You can use the memory keys if you need to multiply twice before you subtract.

You burn 198 calories an hour bowling. You burn 72 calories an hour watching television. How many more calories do you burn bowling for 1.5 hours than watching television?

Press: [1] [9] [8] [×] [1] [.] [5] [M+]

Press: [7] [2] [×] [1] [.] [5] [M−] [MRC] | 189 |

Find how many more calories you burn in 1.5 hours doing the first activity than doing the second activity.

1. Karate, 441 calories an hour
   Volleyball, 207 calories an hour

2. Hiking, 414 calories an hour
   Walking, 297 calories an hour

DECISION MAKING:
How Much of Each Ingredient Is Needed?

Simon is the head chef for the Orlando Café. He went to school for four years to learn how to become a chef. Simon works many hours each day. He enjoys planning his own menus.

Simon is planning a dinner party for 32 people. Simon needs to make sure that he has all of the ingredients for the dinner. The dinner menu includes a salad. The recipe for Simon's special salad dressing is on the right.

Use the recipe above to solve each problem.

1. The recipe serves 4 people. How much of each ingredient is needed to serve 8 people? (Hint: Double the recipe.)

2. Simon needs to serve 32 people. How much of each ingredient will he need?

3. A cup is 8 ounces. How many ounces of olive oil will Simon need to make the dressing for 32 people?

4. A bottle of vinegar is 16 ounces or 2 cups. Will one bottle be enough to make the dressing for 32 people?

SIMON'S SPECIAL DRESSING

Serves 4 people

$\frac{1}{2}$ cup olive oil

$\frac{1}{4}$ cup vinegar

1 teaspoon minced garlic

2 tablespoons thyme

$\frac{1}{2}$ tablespoon marjoram

You Decide

Simon's Special Dressing is so popular that the restaurant decides to sell bottles to customers. It costs about $0.99 to make each bottle of salad dressing. How much should the restaurant charge for each bottle to encourage sales and make a profit?

Summary

You can plan a balanced diet by choosing the correct number of servings from each food group.

You can find the amount of protein a person needs if you know the person's age and weight.

You need enough calories to supply your body with energy, but not more calories than your body is able to burn.

You burn the most calories when you exercise and do other active things.

Eating too much fat, salt, and sugar may cause heart disease, high blood pressure, and weight gain.

Use cross products to solve a proportion.

calorie

diet

food groups

proportion

ratio

serving

unsaturated fat

vegetarian

Vocabulary Review

Complete the sentences with the words from the box.

1. A person who does not eat meat is a ____.

2. A ____ is everything you eat and drink regularly.

3. A unit of energy that comes from food is a ____.

4. An oil such as olive oil, corn oil, or safflower oil that is liquid at room temperature is an ____.

5. A ____ is a comparison of one amount with another amount.

6. A ____ is a portion of food or drink usually eaten at one time.

7. A statement that two ratios are equal is a ____.

8. The food categories used to plan a balanced diet are the ____.

Chapter Quiz

Solve each problem. Use the chart for Exercises 1 and 2.

Number of Servings Needed Each Day			
Food Group	Children 2–6	Children 7–12, Teenage girls	Teenage boys
Bread/Cereal	6	9	11
Meat/Poultry/Fish	5 ounces	6 ounces	7 ounces
Vegetable	3	4	5

1. How many servings from the bread/cereal group does a teenage boy need each day?

2. Who needs to eat more meat daily, teenage girls or teenage boys? How much more?

3. A person 11 to 14 years old needs 0.45 gram of protein per pound of weight. Your 13-year-old brother weighs 110 pounds. How many grams of protein does he need in a day?

4. There are 93 calories in 3.5 ounces of red snapper. About how many calories are there in 8 ounces of that fish?

5. One gram of fat has 9 calories. One cup of rice pilaf has 14 grams of fat and 340 calories. What percent of the calories in one cup of pilaf is from fat? Round to the nearest whole percent.

Maintaining Skills

Solve each proportion.

1. $\dfrac{5}{7} = \dfrac{x}{14}$

2. $\dfrac{3}{x} = \dfrac{24}{16}$

3. $\dfrac{x}{8} = \dfrac{15}{32}$

4. $\dfrac{9}{7} = \dfrac{45}{x}$

5. $\dfrac{11}{3} = \dfrac{55}{x}$

6. $\dfrac{x}{12} = \dfrac{84}{42}$

7. $\dfrac{7}{x} = \dfrac{28}{33}$

8. $\dfrac{13}{5} = \dfrac{x}{16}$

The nutrition labels on food items can help you choose the foods you need. What are some foods you need to stay healthy?

Learning Objectives

LIFE SKILLS

- Read and analyze nutrition labels on food items.
- Find and compare unit prices to determine the better buy.
- Analyze store coupons and manufacturers' coupons.
- Learn how to spend money wisely when buying groceries.

MATH SKILLS

- Add, subtract, multiply, and divide with money.
- Find the unit price.

Words to Know

nutrient	something found in food that you need to grow and be healthy
percent daily value	the amount of a nutrient in a single serving compared to the amount of the nutrient you need each day
ingredients	things that make up a mixture like food
unit price	the price of one item or the price per unit of an item; some examples of units are ounce, pound, and quart
brand	a name given to a group of items manufactured or sold by the same company
nationally advertised brands	brand names of food, personal care items, household items, clothes, or other items that are sold throughout the country and are advertised in the newspapers and on television
store brands	brand names owned by a store or chain of stores
expiration date	the last day a product can be sold; the last day a coupon can be used

Project: Create a Shopping List

Create a shopping list for one week of meals for one person. You have $60 to spend on breakfast, lunch, dinner, and snacks. Identify two food stores and a Web site that sells food. Compare prices and look for coupons.

Answer the following questions.

- What is the total cost of the food?
- What kinds of food were more expensive? Less expensive?
- Which grocery store offered the better prices?
- How did you use coupons to save money?

Compare your shopping list with those of your classmates.

United States law requires a nutrition label on most packaged foods. This label tells you if the food is a good source of a **nutrient**. It also helps you compare the nutritional values of different foods.

This label is from a can of peanuts. It gives you a lot of information. It tells you that the serving size is 1 ounce and that there are 12 one-ounce servings in the can.

The number of calories in a single serving of these peanuts is 170. You can see that 120 of the calories in each serving are from fat.

A nutrition label also lists each nutrient. It tells you how much of each nutrient is in a single serving. It also tells you what **percent daily value** of each nutrient is in each serving. For example, the fat in 1 ounce of Aunt Bea's Peanuts is 21% of the daily amount of fat you need in a 2,000-calorie diet.

There is other useful information on the label. The **ingredients** tell you what is in the food. The ingredients are always listed in order from the greatest amount to the least amount. The first ingredient in this can of peanuts is peanuts. So, you know you are getting what you paid for! There are also smaller amounts of corn oil and salt. Salt is listed last. So you know that there is less salt than oil.

Aunt Bea's Peanuts

Nutrition Facts

Serving Size 1 oz (28g)
Servings per container 12

Amount Per Serving	
Calories 170	Calories From Fat 120
	% Daily Value*
Total Fat 13 g	21%
Saturated Fat 2.5 g	12%
Cholesterol 0 mg	0%
Sodium 45 mg	2%
Total Carbohydrate 7 g	2%
Dietary Fiber 2 g	8%
Sugars 1 g	
Protein 7 g	

Vitamin A 0%	Vitamin C 0%
Calcium 0%	Iron 0%

*Percent Daily Values (DV) are based on a 2,000 calorie diet.

Ingredients: Peanuts, Corn Oil, Salt

Here's a Tip!

Food labels use abbreviations.

milligram (mg)
gram (g)
ounce (oz)

Practice and Apply

Use the nutrition label on page 240 to answer each question.

1. How many servings are in a can of Aunt Bea's Peanuts?

2. What is the serving size of these peanuts? How many grams are in one serving?

3. How many calories are in one serving?

4. How many calories in a serving come from fat?

5. How many milligrams of salt are in one serving? (Hint: Sodium is salt.)

6. Look at the percent daily value listed for sodium. What percentage of a person's daily value of sodium is in one serving of these peanuts?

7. Is there any vitamin A in these peanuts? Explain.

8. How much fiber is in one serving of these peanuts? What is the percent daily value for fiber?

9. How much protein is in 2 servings of these peanuts?

10. Corn oil is listed as the second ingredient. What does this mean?

11. CRITICAL THINKING Suppose your doctor tells you that you need more calcium. How could you use nutrition labels to help you increase the calcium in your diet? Would you eat more peanuts?

Stores sell some grocery items in packages of different quantities. You can compare prices if you know the **unit price** of each item. The unit price is the price of one item, one unit, or one serving.

▸ **EXAMPLE 1**

Here's a Tip!
Food stores place labels near each food item on the shelf. These labels show the name of the item, the price of the item, and the unit price.

A case of 24 small bottles of spring water costs $5.52. What is the unit price?

Divide the price by the number of bottles.
$5.52 ÷ 24 = $0.23

The unit price of one bottle of spring water is $0.23.

Some grocery items are packaged by weight or capacity. In this case, the unit price is the price of one unit of measure, such as an ounce.

▸ **EXAMPLE 2**

Suppose a box of cereal costs $3.84. The box contains 32 ounces of cereal. What is the unit price of the cereal?

Divide the price by the number of ounces.
$3.84 ÷ 32 = $0.12

The unit price of the cereal is $0.12.

Sometimes you will need to round your answer.

▸ **EXAMPLE 3**

A 64-ounce bottle of juice costs $3.99. What is the unit price?

STEP 1 Divide the price by the number of ounces.
$3.99 ÷ 64 = $0.06234375…

STEP 2 Round to the nearest cent.
$0.06234375… is about $0.06

The unit price of the juice is about $0.06.

Skills Practice

Find the unit price. Round to the nearest cent if necessary. Use a calculator if you like.

1. $6 for a box of 5 light bulbs

2. $4.71 for a box of 30 trash bags

3. $2 for a bag of 8 apples

4. $6.50 for a box of 4 bars of soap

5. $3.50 for a package of 2 toothbrushes

6. $12.00 for a package of 3 boxes of fabric softener

7. $1.26 for 18 ounces of detergent

8. $2.40 for a 15-ounce box of cereal

9. $7.88 for 4 pounds of meat

10. $1.90 for 2 pounds of potatoes

11. $7.68 for 24 ounces of floor cleaner

12. $1.60 for 4 ounces of lunch meat

Everyday Problem Solving

Look at advertisements A and B. Solve each problem. Show your work.

1. What is the price per roll for the paper towels in advertisement A?

2. What is the price per roll for the paper towels in advertisement B?

3. Both advertisements show Ultra Strong paper towels. The paper towels are the same quality and each roll is the same size. Which one has the better price for paper towels?

How Can You Save Money by Comparing Prices?

12·3

There are some things you can do to save money on groceries. You can compare prices for the same item sold under different **brand** names. Many stores sell two kinds of brands: **nationally advertised brands** and **store brands.** Store brand items are less expensive.

Consumer Beware!
A less expensive item is not always a good buy. Some paper towels cost very little, but do not work as well as a more expensive brand.

You can also compare prices for the same brand items at nearby stores. Suppose the same item is sold in two different stores, but the size and price are different. You can find the unit price of the item in each store. Then compare the unit prices to find out which store has the better buy.

▶ **EXAMPLE**

Frank's Foods sells a 40-oz jar of peanut butter for $7.45. Cathy's Deli sells an 18-oz jar of the same brand peanut butter for $3.15. Which store has the better buy for this peanut butter?

STEP 1 Find the unit price of the peanut butter in each store. Divide the price by the number of ounces in each jar.
Frank's Foods $7.45 ÷ 40 = $0.18625
Cathy's Deli $3.15 ÷ 18 = $0.175

STEP 2 Compare the digits in each price from left to right.
Frank's Foods $0.1**8**625
Cathy's Deli $0.1**7**5
same ⌐⌐ 8 > 7

STEP 3 Compare the prices.
Frank's Foods   Cathy's Deli
$0.18625   > $0.175
The peanut butter in Cathy's Deli costs less.

Cathy's Deli has the better buy for this peanut butter.

Practice and Apply

 Find which store offers the better buy. Use a calculator if you like.

	Food	Frank's Foods	Cathy's Deli	Which Store Offers the Better Buy?
1.	cereal	$3.36 for 16 ounces	$5.52 for 24 ounces	?
2.	meat	$3.87 for 3 pounds	$6.25 for 5 pounds	?
3.	apples	$3.80 for 10	$0.86 for 4	?
4.	soup	$1.99 for 16 ounces	$3.99 for 24 ounces	?
5.	tomato sauce	$2.49 for 26 ounces	$1.89 for 10.5 ounces	?

 Find which brand costs less per ounce. Use a calculator if you like.

	Food	Nationally Advertised Brand	Store Brand	Which Brand Costs Less?
6.	ketchup	$1.69 for 14 ounces	$1.29 for 12 ounces	?
7.	peanuts	$4.29 for 11.5 ounces	$4.19 for 9.5 ounces	?
8.	chili	$1.89 for 18 ounces	$1.35 for 15 ounces	?
9.	oatmeal	$2.49 for 18 ounces	$1.69 for 24 ounces	?

10. WRITE ABOUT IT The same food can be packaged in small containers or in large containers. Usually, the unit price for the food in the larger container is less. Is it always a good idea to buy the food in the larger container? Explain.

You can save money when you buy groceries if you use coupons. A cashier subtracts the amount of the coupon from the price of the item.

There are two types of coupons. Manufacturers' coupons are for national brand items. They can be used in almost any store. A store coupon can only be used at the store that gives out the coupon.

Sometimes a coupon can only be used when you buy more than one item.

▶ **EXAMPLE 1**

Jerome has a coupon for $0.50 off the price of 3 cans of Sea Fresh tuna fish. The tuna fish sells for $1.19 a can. How much will 3 cans of Sea Fresh tuna fish cost Jerome if he uses the coupon?

STEP 1 Multiply to find the cost of 3 cans without the coupon.
$1.19 × 3 = $3.57

STEP 2 Subtract to find the cost using the coupon.
$3.57 − $0.50 = $3.07

The tuna fish will cost Jerome $3.07 if he uses the coupon.

Sometimes stores double manufacturers' coupons.

▶ **EXAMPLE 2**

Jerome has a manufacturer's coupon that offers $0.35 off a box of Soft Tissues. The box of tissues sells for $2.29 in Frank's Foods. This week, Frank's Foods will double manufacturers' coupons. How much will Jerome pay for a box of Soft Tissues at Frank's Foods this week?

Wordwise
Double means *two times.*

STEP 1 To double the coupon, multiply the amount by 2.
$0.35 × 2 = $0.70

STEP 2 Subtract $0.70 from the price of the tissues.
$2.29 − $0.70 = $1.59

Jerome will pay $1.59 for a box of Soft Tissues at Frank's Foods this week.

Practice and Apply

Find each price when you use the coupon.

	Food	Regular Price	Value of Coupon	Price with Coupon
1.	Cereal	$2.45	$0.50	?
2.	Gravy	$1.29	$0.20	?
3.	Soup	$0.95	$0.15	?

Solve each problem. Show your work.

4. Lynda has a coupon that offers $0.35 off the price of a box of cereal. The cereal sells for $3.29. How much will the box of cereal cost if Lynda uses the coupon?

5. Tony has a coupon for $1.00 off the price of 4 cans of Healthy Time soup. The soup sells for $1.59 a can. How much will 4 cans of soup cost if Tony uses the coupon?

6. Harry has a coupon for $0.65 off the price of 2 tubes of Super Bright toothpaste. The toothpaste sells for $2.59 a tube. How much will 2 tubes of toothpaste cost if Harry uses the coupon?

7. Talisha has a coupon for $0.25 off the price of a box of pasta. The pasta sells for $1.75 at Frank's Foods. This week, Frank's Foods will double coupons. How much will Talisha pay for the pasta with the coupon?

8. Rachel has a coupon for $0.50 off the price of two jars of pasta sauce. The sauce sells for $2.09 at Frank's Foods. This week, Frank's Foods will double coupons. How much will Rachel pay for 2 jars of sauce with the coupon?

9. **CRITICAL THINKING** Find coupons for 5 items you use. Suppose you buy the items using the coupons. How much will you save? How much will you save if the coupons are doubled?

What About Limits on the Use of Coupons?

You need to read coupons carefully. A coupon may have an **expiration date.** You can't use the coupon after its expiration date. Some store coupons can't be used unless you spend a minimum amount of money in the store. Some stores limit the number of items you can buy with a coupon.

Sara's Market

ZINGER
DETERGENT
FREE

12-oz bottle of ZINGER Detergent
with $15 purchase.

Limit 1 per customer
Offer Expires 2/15/03

EXAMPLE

Zack is shopping at Sara's Market on 2/10/03. He has the coupon above. These items are in his shopping cart: eggs for $1.59, milk for $3.98, bread for $1.99, cheese for $4.29, and cereal for $3.69. Can Zack use this coupon?

Consumer Beware!
Pay attention as your groceries are checked out. Make sure you are not overcharged. If you are charged a higher price, some stores will give you the item free.

STEP 1   Look for limits on the coupon.
Zack must buy $15 worth of groceries.
The coupon expires on 2/15/03.

STEP 2   To find out if Zack will spend $15, find the total cost of the items in his cart.
$1.59 + $3.98 + $1.99 + $4.29 + $3.69 = $15.54

STEP 3   Compare the total with the amount of the purchase he needs to make.
$15.54 > $15

STEP 4   Read the coupon to see if it has expired.
2/10/03 comes before 2/15/03. The coupon has not expired.

Zack can use the coupon.

Practice and Apply

Solve each problem. Show your work.

1. Amanda buys 2 cans of Craig's Crushed Tomatoes on 7/5/03. Could she use the coupon on the right? Explain.

Craig's Crushed Tomatoes

SAVE **75¢**

When you buy two (2) 8-oz cans

1 coupon per customer   Offer expires 6/30/03

2. Byron buys 2 cans of Craig's Crushed Tomatoes. The price per can is $1.19. Byron buys the tomatoes on 6/20/03. How much will the 2 cans cost if Byron uses the coupon above?

3. Alex has a coupon for $0.50 off any size of Good Farm cottage cheese. The coupon is only for purchases of 2 containers. Alex buys 2 containers. Each container of cottage cheese costs $1.78. Could Alex use the coupon? How much will he pay for the 2 containers of cottage cheese?

4. A coupon offers a free bag of Monty's Bagels with the purchase of $20 worth of groceries at Monty's grocery store. Mari buys soup for $0.59, milk for $2.19, cereal for $4.29, mayonnaise for $2.29, ham for $7.49, and butter for $4.09. Could Mari use the coupon? How much will she spend on the groceries?

5. **IN YOUR WORLD** Create your own coupon. Decide if it will be a manufacturer's coupon or a store coupon. Write the name and size of the item on the coupon. Show the value of the coupon. Include at least two limits.

There are things you can do to stay within your food budget. Try not to shop when you are hungry. If you are hungry, you may buy more than you need. When you plan your meals for the week, plan to use foods that are on sale. Find out what items are on sale before you make your shopping list. Check advertisements and compare prices from different stores.

Wordwise
Processed foods have food coloring, salt, sugar, and other ingredients added by the manufacturer.

Buy fresh fruits and vegetables when they are in season. Foods are in season when they are growing in your part of the country. Do not buy processed snack foods. They are not healthy and can be expensive.

Compare the prices between brands and sizes. You may want to choose a store brand instead of a nationally advertised brand when it is less expensive and the quality is good. Watch out for displays. Don't let the store influence your choices with displays of foods you don't need.

Bring a calculator with you when you shop. Add the prices as you put food into the cart. This will help you stay within your budget. Pay attention as your groceries are checked out. Make sure you are not overcharged.

▶ **EXAMPLE**

Aaron is a cashier at Value Supermarket. By mistake, he entered the wrong price on his cash register. Instead of $3.90, he entered $39.00. How much should Aaron take off the bill to correct the mistake?

Subtract the correct price from the mistake.

$39.00 − $3.90 = $35.10

Aaron should take off $35.10 from the bill to correct the mistake.

Practice and Apply

Solve each problem. Show your work.

1. Gloria is a cashier at Frank's Foods. She entered the wrong price on her cash register. Instead of entering $2.40, she entered $24.00. How much should Gloria take off the bill to correct the mistake?

2. Abe is a cashier for Sara's Market. He entered the wrong price on his cash register. Instead of entering $4.50, he entered $5.40. How much should he take off the bill to correct the mistake?

3. Martin is a cashier for Frank's Foods. He entered the following items: bananas for $1.29, butter for $2.49, and bread for $1.69. The customer paid with a $20 bill. Martin gave her back $4.53, which was incorrect. How much more should she receive? (Hint: First find the total cost.)

4. **IN YOUR WORLD** Suppose a cashier made a mistake when he or she checked out your groceries at the store. What would you say to the cashier?

Maintaining Skills

Find the better buy.

1. 3 jars for $0.81
   10 jars for $2.80

2. 6 cans for $2.40
   12 cans for $4.80

3. 25 ounces for $1.75
   7 ounces for $0.77

4. 12 bottles for $8.40
   4 bottles for $2.80

Solve each problem. Show your work.

1. Lucille is a cashier at Value Supermarket. She enters $53.10 instead of $5.31 by mistake. How much should Lucille take off the bill to correct the mistake?

2. Min is buying food for a party. She has $35.00. She bought carrots for $9.04, celery for $10.60, cauliflower for $7.90, and broccoli for $5.88. How much money does she have left?

3. Linda needs carrots. She sees a 2-pound bag of carrots for $0.99. Loose carrots sell for $0.61 for 1 pound. Which carrots are the better buy?

4. **OPEN ENDED** You have $8.00 to buy lunch. You will choose among the following items: $3.50 for salad, $1.49 for a bag of pretzels, $3.75 for soup, $5.25 for a tuna sandwich, $1.19 for a bottle of water, $1.49 for a bottle of juice, and $2.49 for fruit salad. Which items will you buy?

Calculator

John wants to buy 2 containers of orange juice that cost $2.79 each. He has a coupon for $0.75 off the price with a purchase of 2 containers. He uses a calculator to find the total cost with a coupon.

Press: [2] [.] [7] [9] [×] [2] [−] [.] [7] [5] [=] ▭ 4.83

Find the cost of the groceries with the coupon.

1. Buy 1 package of chicken for $4.99 with a $0.75 coupon.

2. Buy 2 boxes of crackers for $2.50 each with a $0.60 coupon.

3. Buy 3 cans of vegetables for $0.79 each with a $0.50 coupon.

ON-THE-JOB MATH:
Stock Clerk

Hermán is a 22-year-old stock clerk for a large supermarket. He began working for the store when he was in high school. Someday he would like to become the manager of his own store.

Hermán is responsible for keeping the shelves neat and well stocked. He pays close attention to displays at the end of each aisle. These displays help to get the customers' attention.

Hermán checks that all the prices are correct. He makes sure that the prices in the computers are the same prices shown on the shelves. Hermán also has to put the unit price on a sign below each item.

1. Hermán has to stock the soup aisle. He has room for 840 cans of soup. There are 24 cans of soup in a case. How many cases of soup can he fit on the shelves?

2. Hermán has to mark down the price of cheese by $0.35 per pound. The cheese is marked at $3.89 per pound. What will the new price be?

3. Suppose the price of an 8-oz can of green beans is $0.96. What is the unit price Hermán would put on the sign?

You Decide

Hermán's boss asks him to create a new store display. The theme of the display is packing a healthy lunch for school. What kind of food items should Hermán include in the display? What non-food items might Hermán include in the display?

Chapter 12 ▷ Review

Summary

Food labels give you information about the nutrients in the food, the number of calories, and the ingredients.

Ingredients on a food label are listed in order from greatest to least amount.

You can find the unit price by dividing the price by the number of items or by the number of units.

You can use the unit price to compare brands and prices among stores.

You can use manufacturers' coupons and store coupons to save money.

Some coupons have limits and expiration dates.

You can shop wisely for groceries if you stay within your budget, buy foods that are good for you, and compare prices to find the best buy.

brand

expiration date

ingredients

nationally advertised brands

nutrient

percent daily value

unit price

Vocabulary Review

Complete the sentences with the words from the box.

1. The amount of a nutrient in a single serving compared to the amount of the nutrient you need each day is the _____.

2. The last day a product can be sold or the last day a coupon can be used is the _____.

3. Something found in food that your body needs to grow and be healthy is a _____.

4. A _____ is the price of one item or the price per unit of an item.

5. _____ are things that make up a mixture like food.

6. A _____ is a name given to a group of items manufactured or sold by the same company.

7. Brand names of food sold throughout the country and advertised in the newspaper and on television are _____.

Chapter Quiz

Solve each problem. Show your work.

1. What does the order of ingredients listed on a food label tell you?

2. Frank's Foods sells a 40-oz jar of mayonnaise for $6.59. Cathy's Deli sells an 18-oz jar of the same brand of mayonnaise for $3.24. Which store has the better buy?

3. Jerome has a coupon for $0.35 off the price of a tube of SoClean Toothpaste. The store offers double coupons this week. The toothpaste sells for $2.19. How much will Jerome pay for the toothpaste with the coupon?

4. Jerome uses the coupon on the right. The regular price for a can is $1.19. On 2/15/04 he buys two cans of Whole-Peeled Tomatoes. How much does he pay for the two cans?

1 coupon per customer

Buy TWO (2) 8-oz cans of *Whole-Peeled Tomatoes* and get

50¢ off.

Offer expires 9/30/05

Maintaining Skills

Find each unit price. Round to the nearest cent.

1. $5 for a case of 10 cans of tuna

2. $3.40 for a box of 20 trash bags

3. $1 for a bag of 5 apples

4. $3.30 for a box of 3 bars of soap

5. $5.97 for 3 lb of meat

6. $2.12 for 10 lb of potatoes

7. $4.77 for 9 oz of cheese

8. $2.20 for 5 oz of lunch meat

Unit 6 **Review**

Write the letter of the correct answer.

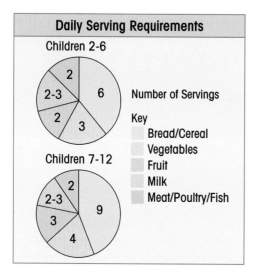

Daily Serving Requirements

Children 2-6

2
2-3 6 Number of Servings
2 3 Key
    Bread/Cereal
Children 7-12 Vegetables
    Fruit
2 Milk
2-3 9 Meat/Poultry/Fish
3
4

1. Ishmael needs 11 servings of bread/cereal a day. His brother Jaron is 8. How many fewer servings of bread/cereal does Jaron need each day?

   A. 2
   B. 3
   C. 5
   D. 9

2. Dora is 5 years old. What is the total number of servings she should eat each day from all food groups?

   A. 12–13
   B. 14–15
   C. 16–17
   D. 18–19

3. Each day Greg needs 0.39 grams of protein per pound of his body weight. He weighs 151 pounds. How much protein does he need every day?

   A. 43.58 grams
   B. 54.99 grams
   C. 58.89 grams
   D. 60.04 grams

4. Joe has a coupon for 75¢ off 2 cans of Silverado brand peas. If each can costs $1.29, how much will he pay if he uses the coupon?

   A. $2.58
   B. $2.17
   C. $1.83
   D. $1.08

5. At Libby's Market a 20-oz bottle of ketchup costs $2.00. What is the unit price?

   A. $0.08
   B. $0.10
   C. $1.00
   D. None of the above

Challenge

One gram of fat has 9 calories. One serving of popcorn has 12 grams of fat. One serving of popcorn has 160 calories. What percent of the calories in popcorn are fat calories?

Chapter 13 **Deciding What You Need**

Chapter 14 **Getting the Best Buy**

When you shop for personal items, you should think about quality as well as price. Visit different stores to learn the best price.

Price of Sneakers

Price: $60, $50, $40, $30, $20, $10, $0

Regular Price · Sale Price

Key ▮ Store A ▮ Store B

The bar graph shows the regular price and sale price of a pair of sneakers. Use the graph to answer each question.

1. What is the regular price of sneakers in Store A?

2. What is the sale price of sneakers in Store B?

3. What is the difference between the regular price of sneakers in Store A and the regular price of sneakers in Store B?

4. What is the difference between the regular price of sneakers in Store B and the sale price of sneakers in Store B?

257

Smart shoppers compare price and quality before they make a purchase. What are some things you look for before you buy?

Learning Objectives

LIFE SKILLS

- Identify things to consider for personal spending.
- Find the sales tax on items.
- Analyze your clothing budget.
- Determine which clothes are right for work.
- Determine the better buy.
- Analyze store advertisements.
- Find the retail price.

MATH SKILLS

- Add, subtract, multiply, and divide with money.
- Find the percent of a number.

Words to Know

appearance	the way something or someone looks
sales tax	money paid to the government based on a percent of the cost of an item
size	how large something is
quantity	the amount or number of something
quality	how well made or well done something is
wholesale price	the amount of money a store pays for an item
retail price	the amount of money a store charges the customer
markup	the difference between the wholesale price and the retail price

Project: Comparing Advertisements

Look at advertisements in the paper and on the Internet. Choose one item that you need to buy. It might be a coat, a household item, sneakers, school supplies, or anything else you need. Find as many advertisements as you can for that item. Compare the advertisements. What information can you find about size, color, quantity, quality, or special offers? Read each ad carefully, especially the small print. Then decide where you will buy the item. Write a paragraph telling why you made that choice.

What Do You Really Need?

Have you ever gone shopping for something you needed, such as socks, and come home with a sweater instead? The sweater looked great, so you bought it without thinking about whether you really needed it or could afford it. If you do this regularly, you might not have money to buy the things you really need. Remember, you must stick to your budget!

Before you spend money on things you want, be sure you have enough money for what you need. Keep in mind three important parts of your life:

• your health

• your **appearance**

• the way you spend your time

You need to save money for doctor and dentist visits, haircuts, clothes, and doing things to keep yourself active and healthy.

Did You Know?
There are different types of health insurance. Some just pay for large medical or hospital bills. Some pay for dental costs, eye care, and even the bills for regular checkups.

If you are working, your employer might offer you health insurance. This will help pay for your medical bills. However, it might not pay for all of your health care.

Your health is most important. Here are some things you need to do:

1. See a doctor if you're sick or injured.

2. See a dentist twice a year to clean and check your teeth.

3. Exercise regularly to keep in good shape.

Taking care of health care issues right away could save you money in the future. You'll miss fewer days of work. Any health problems you do have will often be less serious.

Practice and Apply

Solve each problem. Show your work. Use a calculator if you like.

1. Clark finds that in an average month, he spends $35 on things he really doesn't need and can't afford. About how much does he spend on these items in a year?

2. Rhonda goes shopping for a new pair of shoes. She plans to spend $40. Instead, she buys a shirt for $29 and a pair of pants for $35. How much more did she actually spend than she planned to spend?

3. Wilmer spends $12 on a haircut every 4 weeks. How much is that in one year?

4. Minsoo gets 9 haircuts a year. Each haircut costs her $20, plus a $3 tip. How much should she budget for haircuts each year?

5. Arlo lost a filling in his tooth. It costs $80 to replace a filling. Arlo didn't go to the dentist. Later that month, Arlo's tooth cracked. Now he needs to have the tooth capped. This will cost $400. How much money would Arlo have saved by having the filling fixed right away?

6. **CRITICAL THINKING** Lia needs a prescription allergy medicine. The medicine costs $24 a month. Lia does not buy the medicine this month. She spends the $24 on a new purse instead. Then, Lia has a bad allergy attack. She misses two days of work and has to go to the doctor. She loses $128 in pay, spends $90 for the doctor visit, and spends $24 for medicine. How much more money would Lia have if she had bought the medicine she needed rather than the purse she wanted? (Hint: Think about the price of the purse and the other expenses related to not taking the medicine in the first place.)

You go shopping and see the belt you need to match your new pants. The price of the belt is $17. But, the clerk says you owe $18.02 for your purchase. Why is the price higher? Some states charge **sales tax.**

Sales tax is a percent of the cost of an item. You add sales tax to the price of an item to find the total cost.

▶ **EXAMPLE**

The price of a book is $9.50. The sales tax rate is 6%. What is the total cost of the book?

STEP 1   Change the percent to a decimal.

6% = 0.06

STEP 2   Multiply the cost of the book by the decimal. This gives you the amount of sales tax.

$9.50 × 0.06 = $0.57

STEP 3   Add the sales tax to the cost of the book.
$9.50 + $0.57 = $10.07

The total cost of the book is $10.07.

Practice and Apply

Find the sales tax for each item. Then find the total cost.

	Cost of Item	Sales Tax Rate	Amount of Sales Tax	Total Cost
1.	$75	5%	?	?
2.	$45	6%	?	?
3.	$130	8%	?	?
4.	$660	7.5%	?	?

5. **CRITICAL THINKING** You pay $252 for new school clothes. The price of the clothes was $240. How much did you pay in sales tax? What percent of the price of the clothes is the sales tax?

Which Clothes Do You Need?

What you wear depends on what you do. How do you spend most of your time? You'll need to buy clothes for these activities first. What do you do at other times? You will need clothes for these activities next.

Suppose your weekly schedule includes the following:

- 30 hours at school

- 15 hours at a part-time job in a factory

- 7 hours with friends on the weekend

You spend most of your time at school, so clothes for school are most important. Most of your clothing budget should be spent on school clothes. These clothes can also be worn on weekends, unless you're going somewhere special.

When you shop for clothes, make a list. Stick to it. Otherwise, you might spend money on clothes you don't need.

▶ **EXAMPLE**

You go to school. You also have a part-time job. You buy 2 shirts for $25 each, pants for $19, and 3 packs of socks for $6.50 each. The sales tax is 6%. How much will your new clothes cost?

Did You Know?
Most clothes are considered necessities, or needs. Some states do not tax necessities. So, in these states you can buy your clothes without paying tax.

STEP 1 Multiply to find the cost of both shirts.
$25 × 2 = $50

STEP 2 Multiply to find the cost of all the socks.
$6.50 × 3 = $19.50

STEP 3 Add to find the total cost without tax.
$50 + $19.50 + $19 = $88.50

STEP 4 Multiply to find the sales tax. Change 6% to 0.06.
$88.50 × 0.06 = $5.31

STEP 5 Add to find the cost of the items with sales tax.
$88.50 + $5.31 = $93.81

You will pay $93.81 for your new clothes.

Practice and Apply

 Use the advertisement below to solve each problem. Sales tax on all items is 8%. Round to the nearest cent if necessary. Use a calculator if you like.

STOREWIDE SALE

Women's Sweater	$34.99
Women's Sleeveless Blouse	$27.50
Men's Casual 3-Button Shirt	$19.99
Men's Socks (Two pairs in a package)	$6.99
Men's T-Shirt	$8.50

1. Franco is a student. He also works 15 hours a week. He has plenty of work clothes, except for socks. He buys 2 casual 3-button shirts and 2 packages of socks. How much does he spend? (Hint: Don't forget to add sales tax.)

2. Laura needs summer blouses. She bought 1 blouse and 2 sweaters. How much did she spend? Did she buy clothes that matched her summer needs?

3. Carla buys a sweater and 2 blouses for work. How much does she spend?

4. Jessie has $55. He buys a T-shirt. How many packages of socks can he buy?

5. **WRITE ABOUT IT** Lois needs new socks for gym class. She planned to buy 6 pairs of socks. When Lois got to the mall, she bought a new blouse instead. Then she did not have enough money for the socks. Lois wears a uniform to school. She already has 11 blouses. Was this a wise decision? Why or why not?

Dressing properly for your job is very important. Here are some tips to help you make wise decisions.

- Notice what people wear at work, especially your co-workers. Do they wear jeans or dress pants?

- Buy clothes that are made well. It is better to have a few well-made clothes than many clothes that will wear out quickly or shrink when you wash them.

- Choose clothes that are comfortable. Can you sit, bend, and stretch in your clothes? Think about the temperature where you work. Is there air conditioning? Is there heat? Are you outside much of the time?

- If possible, try to buy clothes you can wear in other places, too. Can you wear your work clothes when you go out with friends? To school? To other places you often go?

- Read the clothing care labels before you buy. Does the item need to be dry-cleaned? You can save money by buying clothes you can wash yourself.

Did You Know?
If you buy clothes late in the season, you can often get a better price. Clothes for summer will be on sale as summer ends. Clothes for winter will be on sale as winter ends.

Practice and Apply

Solve each problem. Show your work.

1. Arnold has $85. He buys khaki pants for $42, a white shirt for $18, and socks for $5.50. Sales tax is 8%. How much money does he have left over?

2. You buy a jacket for $125. It will cost $6 to dry-clean the jacket each time. If you have the jacket cleaned twice a year, how much will it cost you to buy the jacket and keep it clean for 5 years? Use a calculator if you like.

3. **WRITE ABOUT IT** Your boss has just announced that you will wear a uniform to work. Write a letter to a friend explaining how you feel about this decision.

When you think about your appearance, you probably think about clothes first. Other things you need to think about include your hair, teeth, nails, and skin.

Look at the items below. Which ones could be used to maintain your appearance?

   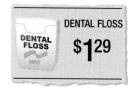

The soap, shampoo, and dental floss help keep you clean, well groomed, and healthy. They are personal care items. The clock is a household item.

▶ **EXAMPLE**

Ivy bought a birthday card for $1.99, a three-pack of soap for $3.15, a candle for $5.07, deodorant for $2.75, and shampoo for $2.70. Sales tax is 5%. How much did she spend on personal care items with tax?

STEP 1 Identify the personal care items.

Soap, deodorant, and shampoo are personal care items.

STEP 2 Add to find the total cost of the personal care items without tax.

$3.15 + $2.75 + $2.70 = $8.60

STEP 3 Multiply to find the tax. Change 5% to 0.05.

$8.60 × 0.05 = $0.43

STEP 4 Add the tax to the total cost of the personal care items.

$8.60 + $0.43 = $9.03

Ivy spent $9.03 on personal care items with tax.

Practice and Apply

 Solve each problem. Round to the nearest cent if necessary. Use a calculator if you like.

1. Edgar buys a package of razors for $8.99 and a can of shaving cream for $3.29. Sales tax is 8%. How much did he pay for these items?

2. Josefina bought a bunch of flowers for $4.99, a bottle of shampoo for $2.29, a tube of hair gel for $3.29, a magazine for $2.95, and a can of deodorant for $1.79. Sales tax is 6%. How much did she spend on personal care items, with tax?

3. Kim likes to keep her fingernails and toenails neat and painted. A manicure and pedicure at a nail salon cost $28 altogether. Kim spent $30 for a home manicure kit and $20 for enough nail polish to last a year. If Kim gives herself a manicure and pedicure every two weeks, how much will she save in one year?

4. Sal likes his dress shirts ironed. The cleaners charge $1.29 to iron each shirt. Sal brings 20 shirts to the cleaners each month. Sal can buy an iron and an ironing board for $50. How much money would he save after three months if he ironed the shirts himself?

5. **CRITICAL THINKING** Ben gets his hair cut once a month for $20. He also gives a 20% tip. His friend offers to cut Ben's hair once a month for half the total amount. How much would Ben save in one year if his friend cuts Ben's hair?

Maintaining Skills

Find the unit price of each item. Round to the nearest cent.

1. 24 oz of shampoo for $3.29

2. 12 oz of lotion for $4.89

3. 8 oz of nail polish remover for $1.29

4. 10 oz of shaving cream for $1.99

13·6 ▷ Is a Sale Item Always a Better Buy?

How do you know if an item on sale is really a good deal? When you see a price that is lower than usual, find out how the item compares with items that are not on sale. Ask yourself these questions:

- What is the **size**? The item might cost less because it is smaller.

- What is the **quantity**? The item might cost less because there are fewer in the package.

- What is the **quality**? The item might cost less because the product is not well made and will not last very long.

Here's a Tip!
Don't buy items you don't need just because they are on sale.

You need to look at size, quantity, and quality to compare the cost of two items.

▶ **EXAMPLE**

Talia buys a package of three rolls of camera film on sale for $8.97. The film is a brand she knows and a good quality. Her friend buys one roll of the same film for $3.79 in a different store. Who got a better buy? Explain.

STEP 1 Find the unit price for the film Talia bought.
$8.97 \div 3 = \$2.99$

STEP 2 Compare the unit price that Talia paid to the unit price that her friend paid for the same film.
$\$2.99 < \$3.79$

Talia got the better buy. She paid the lower unit price.

However, if Talia doesn't take many pictures, it may make more sense to buy only one roll. Even though the unit price is higher, Talia will spend less money than if she had bought three rolls.

Practice and Apply

Solve each problem. Show your work.

1. A box of 12 bags of microwave popcorn is on sale for $8.28. How much does the popcorn cost per bag?

2. A brand-name shirt costs $19.95. A similar shirt costs $15.95. You notice that the less expensive shirt has a few loose buttons. Which shirt is the better buy? Explain.

3. Batteries cost $3.99 for a package of 4, or $2.29 for a package of 2 of the same brand. Which is the better buy?

4. A 2-ounce breakfast bar costs $0.79. A 1.5-ounce breakfast bar of the same brand costs $0.69. Which is the better buy?

5. One undershirt sells for $6.50. The same undershirt is sold in a package of 3. The package costs $14.25. If you plan to buy 3 shirts, how much do you save by buying the package?

6. **WRITE ABOUT IT** Write a letter to a store or manufacturing company, telling them about an item you purchased that wore out too quickly or broke. Include details about size, quantity, and quality that could have made the item better.

Maintaining Skills

Divide. Round to the nearest cent.

1. $4.89 \div 4$
2. $3\overline{)\$6.50}$
3. $8.50 \div 12$
4. $12\overline{)\$144}$

5. $2\overline{)\$0.89}$
6. $15.99 \div 6$
7. $6\overline{)\$18.50}$
8. $11.49 \div 5$

Advertisements, or ads, can be tricky. Sometimes they don't give you all the information you need. Sometimes they include information in very small print or in a different part of the ad. Look at this ad.

Digital Camcorder/Camera
- 22x optical/44x digital zoom
- 8MB memory card for still images
- 2.5-inch color LCD screen

$1,200

*only $68/month***

*With a Camera City credit card. Subject to credit approval.

The ad says you can pay $68 a month for the camera. However, next to the offer, you see an asterisk (*). This means you should look at the bottom of the ad. There you will find a sentence with another asterisk in front of it. It says you need a store credit card and a good credit rating to buy the camera for this price.

Sometimes a store will offer an item at a low price if you purchase another item. At other times, you need to buy more than one of the same item to receive the advertised low price. For example, a store might sell a shirt for $14 if you buy two or more of them. If you only buy one, the shirt might cost $19. These are just some of the strategies used when stores advertise.

When you read advertisements, check to see if the price includes any of these extra costs:

- Delivery charges, especially for large items

- Assembly charges for putting the parts together

- Installation charges for hooking up an item

Consumer Beware!
When an item is offered for a monthly price, there are almost always additional interest costs. This means that you can make a payment each month on the cost of the item, but it will cost you more than the actual purchase price.

Practice and Apply

Use the advertisements below to solve each problem.
Show your work.

Light Bulbs

Buy one 2-pack,
get the second FREE*

*Saturday and Sunday only

Videotapes

3-Pack of 6-hour tapes

$2⁹⁹* plus tax

*With the purchase of any VCR.
Regular price of video tape, $5.99 plus tax.

Personal Computer
ONLY $599*

1.2 GHz Processor
128MB PC2100 DDR SDRAM
20GB Hard Drive
48X CD-ROM

*With any 2-year Internet service contract

1. Lee wants to get a free pack of light bulbs. What does he have to buy?

2. Ivana bought the personal computer. Sales tax on the computer is 5%. The Internet service plan that she must sign up for is $24.99 a month, including tax. What is the total cost of the computer and the Internet service at the end of the Internet service contract?

3. Sarie purchases a pack of videotapes. She gives the sales clerk $10. She gets $3.65 in change. Sarie tells the clerk that this is incorrect. The sales clerk tells her the price is $5.99 plus tax. Sarie says she saw a sign that says $2.99 plus tax. Who is correct? Why?

4. **IN YOUR WORLD** Find an advertisement you have seen in the newspaper, in a mailing, or in a store that could be misleading. Explain why.

Store owners want you to shop at their stores. They advertise items so people will want to buy them.

One way they do this is by offering items at wholesale prices. The **wholesale price** is what the store pays for an item. The **retail price** is what a customer usually pays for an item. The retail price is the wholesale price plus a **markup**.

Look at this advertisement.

Consumer Beware!
When you are shopping during a sale, pay close attention to the signs. A sign that says 70% off may mean 70% off the sale price or it may mean 70% off the retail price. There might be a big difference in these two prices.

SKI WEAR
BELOW Wholesale Prices!
UP TO **70%** *OFF!*
SUMMER SALE

Notice that the store is offering to sell ski wear below wholesale prices. If you owned a store, would you sell an item for less than you paid for it? Probably not. If you did, you would not make money on that item. The prices of the ski wear may be low, but they're probably not below what the store paid.

▶ **EXAMPLE**

A store owner buys a case of 12 cans of tennis balls. The wholesale price of the case is $10.99. The markup is 80%. What is the retail price of the case?

STEP 1 Change the percent to a decimal.
80% = 0.80

STEP 2 Multiply the decimal by the wholesale price to find the amount of the markup. Round to the nearest cent.
$10.99 × 0.80 = $8.792 → $8.79

STEP 3 Add the markup to the wholesale price.
$8.79 + $10.99 = $19.78

The retail price of the case of tennis balls is $19.78.

Practice and Apply

Solve each problem. Show your work. Round to the nearest cent if necessary.

1. Carly owns a T-shirt shop. She pays $2.25 for each shirt. Carly prints a picture on each shirt. The printing costs her $1.00 per shirt. She sells the shirts for $10.50 each. How much is the markup on each shirt?

2. Jason owns a snack bar. He buys a case of 24 bottles of juice for $9.79. He sells each bottle for $0.75. How much is the markup on one case of juice?

3. Gina works at a clothing store. She buys scarves from the warehouse for $12 a scarf and marks up the price by 75%. How much does she charge customers for 1 scarf?

4. Garcia owns a hair salon. He buys shampoo from a warehouse for $2.50 a bottle. He marks up the price of each bottle by 70%. How much does Garcia charge per bottle?

5. Laurel works at a children's store. She buys blankets from a warehouse for $8.50 each. She marks up the price of each blanket by 60%. How much does Laurel charge per blanket?

6. **CRITICAL THINKING** Ashton sells kitchenware. He marks all the frying pans down from $24 to $18. Then he puts a sign up that says, "Frying Pans: 25% off." A customer brings a frying pan to the register and is surprised when Ashton tells him the price is $18. The customer says the price should be 25% off from the labeled price of $18. Who do you think is correct? Why?

Solve each problem. Show your work.

1. Gregg spends $18 every 4 weeks for a haircut. How much will his haircuts cost him in one year?

2. David's yearly clothing budget is $1,400. He buys a suit for $250, 2 shirts for $24 each, a tie for $18, and shoes for $49. Sales tax is 5%. How much does he have left in his yearly clothing budget?

3. Anila bought a 3-pack of air fresheners for $3.99. One air freshener costs $1.39. How much did she save by buying the 3-pack?

4. **OPEN ENDED** Gloria has $85 to spend on clothes for work and for recreation. From this $85, she needs to spend from $20 to $40 on clothes for recreation. How much could she plan to spend on work clothes?

Calculator

You can use a calculator to find the cost of an item including tax.

You buy a gift for $27. The sales tax rate is 6%. So, the total cost of the gift is the price of the gift + 6% of that price.

Press: [2] [7] [+] [6] [%] | 28.62 |

Find the total cost of each item, including tax. Round to the nearest cent.

	Cost of Item	Sales Tax Rate	Total Cost
1.	$52.00	5%	?
2.	$78.50	6%	?
3.	$64.65	7%	?
4.	$88.75	2.5%	?

DECISION MAKING:
Which Clothes Best Fit Your Needs?

Tom needs to buy some new clothes for work. He is a production assistant for a big company.

He works in a large office building with about 300 other employees. He spends a lot of his time at his desk working on his computer. He also has weekly meetings in the office.

Tom's boss wears dress pants, a button-down shirt, and a tie. Some people wear sweaters.

Here are the items Tom is thinking about buying to wear to work.

Use the chart and the information above to answer each question. Tom can only spend $85.

1. Which item does Tom not have enough money to buy?

2. Which two items would be considered inappropriate for work?

3. Tom wants to buy the wool pants. What should he buy to wear with the pants? Why?

Item	Cost
Gray sport jacket	$72
Black wool pants	$52
Button-down shirt	$28
Jeans	$40
Sneakers	$38
Cotton sweater	$39
Silk sweater	$98
Black shoes	$54

You Decide

Tom received some money for his birthday. He has $150 to spend. What items should he buy? Explain your choices.

Chapter **13** Review

Summary

Your health, your appearance, and the way you spend your time are important things to consider for personal spending.

To find the amount of sales tax, multiply the sales tax rate by the cost of the item.

The clothes you spend the most money on should be the ones you'll wear most often.

It is important to wear appropriate and comfortable clothes to work.

You need to think about size, quantity, and quality to find the better buy.

Finding the unit cost can help you find the better buy.

Some advertisements are misleading. You need to read the fine print and be aware of extra costs.

To find the retail price, add the markup to the wholesale price.

appearance

markup

quality

quantity

retail price

sales tax

size

wholesale price

Vocabulary Review

Complete the sentences with the words from the box.

1. The amount of money a store charges a customer is the _____.

2. _____ is the way something or someone looks.

3. The amount of money a store pays for an item is the _____.

4. The difference between the wholesale price and the retail price is the _____.

5. _____ is how well made or well done something is.

6. _____ is how large something is.

7. The amount or number of something is the _____.

8. Money paid to the government based on a percent of the cost of an item is _____.

Chapter Quiz

Solve each problem. Round to the nearest cent if necessary.

1. Brendan buys clothes for basketball. He buys 3 T-shirts that are
   $9.50 each, a new pair of sneakers for $59.50, and 2 pairs of shorts
   that are $19.50 each. Sales tax is 8%. What is his total cost?

2. Jacyln buys the following: body lotion for $4.25, body wash for
   $2.50, a candle for $5.50, dental floss for $1.75, and a notebook
   for $3.00. Sales tax is 4%. How much did she spend on personal
   care items, including sales tax?

3. Molly's Shop sells a package of three rolls of paper towels for $3.79.
   At Great Shop you can buy two of the same rolls at $1.49 each and
   get one free. Which store has the better deal? What is the price of
   one roll of paper towels there?

4. Lily buys a new computer advertised for $799. Sales tax is 5%. She
   must also pay for 1 year of Internet service for $15.99 a month,
   including tax. What is the total cost?

5. Morgan bakes pies for a bake sale. Each pie costs $2 to make. She
   marks up all pies by 75%. How much does Morgan charge for
   each pie?

Maintaining Skills

Multiply.

1. $18 × 0.06   2. $28 × 0.08   3. $42 × 0.05   4. $35 × 0.07

Add.

5. $18 + $1.08   6. $28 + $2.24   7. $42 + $2.10   8. $35 + $2.45

You can shop in a store, from a catalog, or online. Look for sales. What are the benefits of shopping online?

Learning Objectives

LIFE SKILLS

- Determine the best times of year to find sales.
- Find the discount.
- Find the sale price.
- Complete an order form.
- Find the total cost of a catalog or online purchase, including tax and shipping costs.

MATH SKILLS

- Find the percent of a number.
- Add, subtract, multiply, and divide with money.

Words to Know

catalog	a listing of products you can order and have sent to you
online	on the Internet
clearance sale	a markdown of prices in a store to reduce stock
irregular	not perfect; the result of a mistake in manufacturing
factory outlet	a store that sells items directly from the factory at a reduced price
sale price	the cost of an item that has been marked down

Project: Selling a Watch

Create an advertisement to sell a watch. This advertisement will go in a catalog. In your advertisement, include:

- a description of the watch
- the regular price
- a discount offering 40% off the regular price
- the date the discount starts and ends
- the sale price
- a drawing or picture of the watch
- the cost for shipping and handling

Be sure the description of the watch includes the color and brand name and whether it is for men or women.

You can get ideas for your own advertisement by looking on the Internet or in catalogs.

When Are Items on Sale?

There are many ways to shop. You can visit a store. You can also buy from a **catalog** or shop **online**. These choices allow you to shop when it's convenient for you.

To save money, buy items on sale. During a sale, you get a discount off the retail price. Sometimes a store has a **clearance sale**. At this time, prices are very low.

Did You Know?
Clearance sales usually start at the end of the season or after a holiday.

Some items are on sale during certain times of the year. The Sale Calendar below shows when some items are usually on sale.

Sale Calendar

January	February	March	April	May	June
Home Goods Clothes Jewelry Shoes	Jewelry Coats Cosmetics	Clothes Shoes	Toys Clothes Shoes Coats Home Goods	Clothes Shoes Jewelry Cosmetics Home Goods	Clothes Shoes Home Goods

July	August	September	October	November	December
Home Goods	Clothes Swimsuits Shoes Home Goods	Clothes Shoes Coats Swimsuits	Cosmetics Home Goods	Toys Home Goods Jewelry Clothes Shoes	Cosmetics Home Goods Jewelry Clothes Shoes Toys

Whenever you shop, always count your change.

▶ **EXAMPLE**

Wordwise
Sheets, comforters, towels, and tablecloths are examples of *home goods*.

Dan buys towels for $32.50 on sale. He pays with two $20 bills. How much change will he receive?

Subtract the cost of the purchase from the value of the bills.

$40.00 − $32.50 = $7.50

Dan will receive $7.50 in change.

Use the Sale Calendar on page 280 to answer each question.

1. During which two months are swimsuits on sale?

2. During which three months are coats on sale?

3. During which three months are toys on sale?

4. What three items are on sale for nine months of the year?

Solve each problem. Show your work.

5. Dan buys a coat on sale. The total cost of the coat is $68.75 during the sale. Dan gives the salesperson one $50 bill and one $20 bill. How much change will Dan receive?

6. Zack buys a pair of shoes on sale. The total cost of the shoes is $50.15 during the sale. Zack gives the salesperson one $20 bill, three $10 bills, and two dimes. How much change will Zack receive?

7. Marianne buys home goods on sale for $44.75. She gives the salesperson one $50 bill. How much change will she receive?

8. **IN YOUR WORLD** What items do you buy on sale? What time of the year do you buy them?

Maintaining Skills

Find each unit cost. Round to the nearest cent.

1. A box of 4 light bulbs for $3.00

2. A box of 20 trash bags for $3.60

3. A bag of 5 oranges for $2.00

4. 8 ounces of cheese for $3.25

5. A 14-oz box of cereal for $3.80

6. 2 pounds of apples for $2.99

Sometimes a factory makes too many of one item or makes items that are not perfect. The items that are not perfect are labeled **irregular,** even if they look perfect. Such items can be sold at a **factory outlet.** Factory outlets are stores that offer discounts all year long. A factory outlet sometimes sells items for 50% or more off the retail price.

► **EXAMPLE**

Jay found a coat at a factory outlet for 45% off the retail price of $275. How much is the discount?

STEP 1 Change the percent to a decimal.

$45\% = 0.45$

STEP 2 Multiply the retail price by the decimal.

$\$275 \times 0.45 = \$123.75$

The discount on the coat is $123.75 off the retail price.

Practice and Apply

Find each discount.

	Item	Retail Price	Percent of Discount	Amount of Discount
1.	Shirt	$35	50%	?
2.	Pants	$80	40%	?
3.	Shoes	$75	25%	?
4.	Jacket	$110	50%	?
5.	Suit	$250	60%	?

6. CRITICAL THINKING Vanessa found the same pair of shoes at two stores. The retail price at Jan's Shoes is $68 with an additional 40% off. The retail price at Shoe Gallery is $90 with an additional 50% off. Which store offers the better discount? Is that the better buy? Explain.

You learned that stores often put items on sale. During a sale, the retail price of an item is reduced. To calculate the **sale price**, you must know the retail price and the discount. The sale price is what you pay after the discount is subtracted from the retail price.

▶ **EXAMPLE**

Vanessa wants to buy a suit. The retail price of the suit is $134. The suit is on sale for 40% off. What is the sale price of the suit?

STEP 1 Change the percent to a decimal.   40% = 0.40

STEP 2 Find the discount. Multiply the retail price by the decimal. $134 × 0.40 = $53.60

STEP 3 Subtract the discount from the retail price. $134.00 − $53.60 = $80.40

The sale price is $80.40 for the suit.

Practice and Apply

Find each discount and each sale price.

	Retail Price	Percent of Discount	Amount of Discount	Sale Price
1.	$850	50%	?	?
2.	$45	40%	?	?
3.	$300	30%	?	?
4.	$20	40%	?	?
5.	$250	60%	?	?
6.	$650	30%	?	?

7. WRITE ABOUT IT You know that a store will be having a sale. Why would visiting the store before the sale be a good idea?

Almost any item you want can be found in a catalog or online. How do you place an order to buy an item?

You fill out an order form. Then you mail, e-mail, or fax the completed order form to the company. You also send a check or credit card number to pay for the item. The item is then shipped to you.

Most order forms ask for the item name, item number, color, size, quantity, and unit price. Sometimes you will see abbreviations on the order form. The abbreviation *Qty.* means *quantity*. Each item that has its own item number should be on a different line of the order form.

Kelly wants to buy two blue shirts in size medium from a catalog. She found the items on page 20 of the catalog. The item number for the blue shirt is 6-48J. The shirts have a retail price of $9.95 each. She begins to fill out the order form below.

Ship To

Name: Kelly Graham   **Address:** 364 Colonial Avenue

City: Lafayette   **State:** CT   **Zip Code:** 06999

Day Phone Number: 203-555-0123   **Evening Phone Number:** 203-555-0006

Are the billing address and the shipping address the same? Yes

Item Number	Catalog Page	Color	Item Size	Description	Quantity	Unit Price	Total Price of Items
6-48J	20	Blue	M	Long sleeve blue shirt	2	$9.95	$19.90

Practice and Apply

Solve each problem. Use the order form on page 284.

1. Kelly also wants to buy 3 pairs of tan pants. The retail price of the tan pants is $14.95 each. What would Kelly write for the Unit Price on the order form? For the Total Price of Items?

2. Jan wants to buy 4 shirts from the catalog. She wants 2 blue shirts, 1 red shirt, and 1 white shirt. Each color has a different item number. How many item lines will Jan use on the order form for the 4 shirts? Why?

3. Dee wants to buy 3 pens from the catalog. The retail price is $15.99 for each pen. The pens are on sale. If you buy 3 or more pens, the price is reduced to $11.50 each. What will Dee write for the Unit Price?

4. Vanessa wants to buy 2 skirts from the catalog. The retail price is $29.00 for each skirt. The sale price is $15.95 each if you buy 2 or more skirts. What is the total price of the items? (Hint: Multiply the sale price by the number of skirts.)

5. Ivan wants to buy 4 CDs from the catalog. The retail price is $12.95 for each CD. The sale price is $8.95 each if you buy 3 or more CDs. How much will Ivan save on each CD by buying 4 CDs?

6. You want to buy 3 shirts from the catalog. The retail price is $9.95 each. The sale price is $5.95 each if you buy 2 or more shirts. How much would you save altogether by buying three shirts?

7. **WRITE ABOUT IT** You can also place an order by phone. You will need to have some information from the catalog ready before you call. What information should you have?

14·5 How Do You Find the Total Cost of Your Order?

When you place an order, you may need to pay a shipping charge. This is the amount the company charges for sending you the order.

The total price may also include sales tax, depending on where you live. Sometimes the tax is calculated on the cost of the item and the cost of shipping.

EXAMPLE

Did You Know?
Shipping costs sometimes include a handling charge. This covers the cost of packing the item to be sent.

Kelly places an order for one video game and one DVD. The retail price is $11.50 for the game and $19.50 for the DVD. The chart below shows shipping costs. The sales tax is 6% on the total order, including shipping. What is the total cost of Kelly's order, including shipping and tax?

Total Cost of Items	Shipping Cost
Up to $30.00	Free
$30.01 to $90.00	$5.00
Over $90.00	$7.00

STEP 1 Add to find the total cost of the items.
$11.50 + $19.50 = $31.00

STEP 2 Use the chart to find the shipping cost.
$31.00 is between $30.01 and $90.00.
So, $5.00 is the shipping cost.

STEP 3 Add the total cost of the items and the shipping cost.
$31.00 + $5.00 = $36.00

Here's a Tip!
6% = 0.06

STEP 4 Find the sales tax. Multiply the tax rate by the total cost of the items with shipping.
$36.00 × 0.06 = $2.16

STEP 5 Add the tax to the total cost of the items with shipping.
$36.00 + $2.16 = $38.16

The total cost is $38.16 for Kelly's order.

Practice and Apply

Use the chart below to solve each problem.

Total Cost of Items	Shipping Cost
Up to $25.00	$2.50
$25.01 to $50.00	$5.00
$50.01 to $75.00	$7.50
$75.00 and more	$10.00

1. Dianna orders 2 sweaters and 1 coat. The retail price is $25 for each sweater and $75 for the coat. How much is the shipping cost for Dianna's order?

2. Jodi places an order for 3 sweaters. The retail price is $29 each or $19 each if you buy 2 or more. How much is the shipping cost for Jodi's order? (Hint: First, find how much each sweater will cost Jodi.)

3. You place an order for 1 radio and 1 book. The retail price is $26 for the radio and $9 for the book. The company will charge you 6% sales tax on the items only. How much is the tax on your order?

4. Tarek places an order for 1 CD and 1 computer software package. The retail price is $12 for the CD and $26 for the software package. The sales tax is 3% on the total cost with shipping. What is the total cost of Tarek's order, including shipping and tax?

5. Margot places an order for 1 pair of boots, 3 pairs of socks, and 1 scarf. The boots cost $54, the socks cost $5 for each pair, and the scarf costs $19. The sales tax is 5% on the total cost with shipping. What is the total cost of Margot's order, including shipping and tax?

6. **IN YOUR WORLD** You are buying a new winter coat. It costs $115. You can purchase the coat at the store, but it is 30 minutes away. You have a coupon for 10% off the coat if you buy from the catalog. The shipping cost is $15. Sales tax on the coat alone is 5% if you go to the store or order from the catalog. How will you buy the coat?

Problem Solving

Solve each problem. Show your work.

1. A jacket has a retail price of $60. You see the jacket at a factory outlet for 65% off the retail price. How much will it cost at the factory outlet?

2. The price of a sweater is $42 at a store in your town. You can buy the same sweater online for 15% less. You will have to pay $4.95 for shipping if you buy the sweater online. Which is the better buy?

3. The price of a tent in a catalog is $120. The shipping charge is 5% of the price. The sales tax is 6% on the total cost including shipping. What is the total cost of the tent including all charges?

4. **OPEN ENDED** You bought jeans and a shirt that were on sale for 50% off. If you paid $45 total for both the jeans and the shirt, what was the original price of each item?

Calculator

The retail price of a sweater is $50. The sweater is on sale for 30% off. You can use a calculator to find the discount and the sale price. First you find the sale price. Then you find the discount.

Press: [5] [0] [−] [3] [0] [%] | 15 | ← discount of $15

Press: [=] | ⁻   35 | ← sale price of $35

Find each sale price and discount.

	Retail Price	Percent of Discount	Amount of Discount	Sale Price
1.	$52	50%	?	?
2.	$80	65%	?	?
3.	$65	20%	?	?
4.	$88	25%	?	?

ON-THE-JOB MATH:
Customer Service Representative

Mike is a 21-year-old customer service representative for Bright Days Clothing. His job is to take catalog orders by phone and enter the order information into a computer. Mike needs good telephone manners. It is important that he meet each customer's needs.

Mike needs to be an accurate and fast typist. He must have good math skills to make sure he does not make a mistake. He also must be able to explain every charge to the customer.

Bright Days Clothing has a training program for customer service representatives. The company may record some calls to make sure the customer service representative handles the customer's needs correctly.

1. Mike took an order for a $38 shirt, a $25 pair of pants, and a $12 hat. The shipping charge is 5% of the total cost of the items without tax. How much is the shipping charge for this order?

2. The total cost of one order is $80. The shipping charge is 5% of the total cost. A 7% sales tax is computed on the final cost, including shipping. How much is the tax?

3. A customer places an order for items that cost $260. The shipping charge is 5% of the total cost of the items. The sales tax is 7% of the total, including shipping. What is the total cost of the order?

You Decide

Mike has a customer on the phone who is very unhappy. An item that he wants to order is not in stock. How could Mike make the customer happier with the company? Explain what he might say and do in this situation.

Summary

Stores may offer the best sales at the end of a season or after a holiday.
To find the amount of the discount, multiply the percent of the discount by the retail price.
To find the sale price, subtract the amount of the discount from the retail price.
You need to fill out an order form to buy items from a catalog or online.
When you place an order with a catalog or online, you may need to pay a shipping cost.
The shipping cost often depends on the total price of the items ordered.
Sales tax is sometimes added to the cost of an item, depending on where you live.

catalog

clearance sale

factory outlet

irregular

online

sale price

Vocabulary Review

Complete the sentences with words from the box.

1. Information that is available on the Internet is _____.

2. An item that is not perfect or is the result of a mistake in manufacturing is _____.

3. The cost of an item that has been marked down is the _____.

4. A _____ is a store that sells items directly from the factory at a reduced price.

5. A _____ is a listing of products you can order and have sent to you.

6. A _____ happens when a store marks down prices to reduce stock.

Chapter Quiz

Solve each problem. Show your work.

1. Marcia buys toys for $22.75 on sale. She gives the salesperson two $10 bills and one $5 dollar bill. How much change will Marcia receive?

2. You find a television at a factory outlet for 30% off the retail price of $320. How much is the discount?

3. You want to buy a camera. The retail price is $75. The camera is on sale for 15% off. What is the sale price of the camera?

4. The retail price of a shirt is $28. The sale price is $22 if you buy three or more shirts. Evan buys four shirts. What is the total price of the items?

5. You place an order for a pair of pants, two shirts, and a sweater. The pants cost $26.50, the shirts cost $18 each, and the sweater costs $42.50. The shipping charge is 5% of the total order. Sales tax is 4% on the total order, including shipping. What is the total cost of your order with shipping and tax?

Maintaining Skills

Find the unit cost. Round to the nearest cent.

1. $1.77 for a box of 3 cans of tuna   2. $2.88 for 2 pounds of meat

Solve each proportion.

3. $\frac{3}{5} = \frac{x}{10}$ 4. $\frac{2}{x} = \frac{18}{27}$ 5. $\frac{x}{7} = \frac{27}{21}$ 6. $\frac{15}{18} = \frac{5}{x}$

Unit 7 **Review**

Write the letter of the correct answer. Use the bar graph to answer questions 1 and 2.

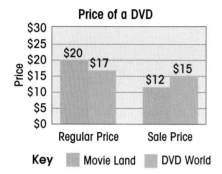

Price of a DVD

Key   Movie Land   DVD World

1. How much does it cost to buy 3 DVDs at the regular price at Movie Land?

   A. $60
   B. $51
   C. $45
   D. $36

2. How much less does it cost to buy one DVD on sale at Movie Land than to buy one DVD at the regular price at DVD World?

   A. $3
   B. $5
   C. $2
   D. $6

3. Trina buys a television for $189. The sales tax is 5%. What is the total cost of the television with tax?

   A. $189.45
   B. $198.45
   C. $9.45
   D. $179.55

4. Lacey's Pro Shop buys a set of golf clubs for $240. The markup on the clubs is 125%. What is the retail price of the golf clubs?

   A. $540
   B. $300
   C. $60
   D. $450

5. Orlando buys curtains for his apartment. The regular price of the curtains is $32. They are on sale for 20% off. How much does Orlando pay for the curtains?

   A. $3.20
   B. $6.40
   C. $25.60
   D. $26.50

6. Eileen buys a frying pan. The regular price is $22. The frying pan is on sale for 35% off. How much money does Eileen save?

   A. $14.30
   B. $6.60
   C. $7.00
   D. $7.70

Challenge

Raj buys a picture frame from a catalog. The regular price is $25. He has a coupon for 20% off. Shipping costs $3. A 6% tax is charged on the sale price and the shipping. What is the total cost of Raj's order?

Unit 8 ▶ Owning a Vehicle

When you look for a car, think about the gas mileage. Gas mileage tells you the number of miles a car will travel on one gallon of gasoline, in the city and in the country.

The bar graph shows the number of miles each car can travel on one gallon of gas. Use the graph to answer each question.

1. How many miles can Car A travel on one gallon of gas in the city?

2. How many miles can Car B travel on one gallon of gas on the highway?

3. Which car uses less gas for both city and highway driving?

293

Chapter 15

Buying and Leasing a Vehicle

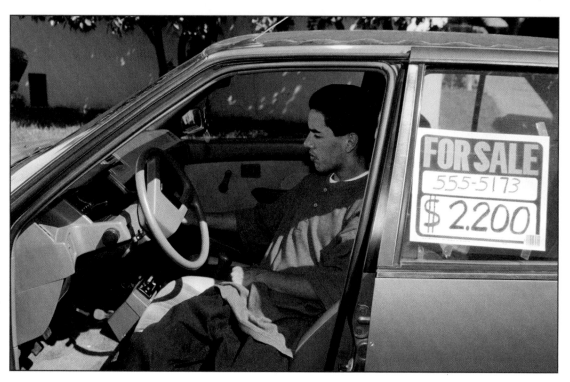

There are many choices you can make when you buy or lease a car. What are some of the things you would want to know before making a decision?

Learning Objectives

LIFE SKILLS

- Choose the best vehicle for your needs.
- Use car price guides to compare car prices.
- Identify ways to buy or lease a car.
- Understand a loan contract.
- Find the monthly car payment amount and the total amount of the car loan.
- Find the residual value of a car.
- Find the charges in a lease agreement.

MATH SKILLS

- Add, subtract, and multiply with money.
- Find the percent of a number.

Words to Know

car price guide	a book that shows the wholesale and retail prices of cars
invoice price	the price the car dealer pays the manufacturer for a car
manufacturer's suggested retail price (MSRP)	the price suggested by the manufacturer for a customer to pay for a new car
warranty	an agreement that the company will replace an item, fix it for free, or return your money if the item cannot be fixed
down payment	money you give the car dealer that lowers the amount left to pay on the car
trade-in	a used vehicle that a car dealer takes from a customer in part payment for another vehicle
lease	to rent a vehicle for a certain period of time; the written agreement to rent
residual value	the value of a car at the end of a lease
depreciation	a decrease in the value of an item because of use and age

Project: Buying or Leasing a Vehicle

Choose two vehicles you would like to own or lease. Find information about both vehicles, including price and gas mileage, on the Internet or in newspapers. Then answer these questions:

- Would you rather buy or lease your vehicle?
- Which equipment and safety features are most important to you? Which are least important to you?
- Which vehicle best fits your budget?

Based on your research, which vehicle would you choose? Write a short report explaining your reasoning. Include a picture of the vehicle with your report.

15·1 What Kind of Transportation Do You Need?

You have a good job, a steady income, and a place to live. You must travel to work or school. What are your transportation needs?

Is your job or school close to home? Could you walk or ride a bicycle to get there? This would keep your transportation costs low.

Could you share a ride with a co-worker or friend? Can you use public transportation? Usually the cost of public transportation is reasonable. Buying a monthly pass saves money.

If none of these options works for you, then you might need a car. It's fun to dream about the car you want most. However, you need to be realistic, too. You should find a car that matches your lifestyle and your income.

Wordwise:
A *stick shift* is also called a *standard transmission* or *manual transmission*.

Think about these things when you look at cars:

• Can you drive a car that has a stick shift, or do you need an automatic transmission?

• How much can you spend on a car each month? The cost of monthly car payments, car insurance, gas, maintenance, and repairs can add up quickly.

• How safe is the car you want? Insurance companies offer discounts for cars with safety features.

• Will you need your car for long trips? If so, the car must not be too old or have too many miles on it.

• Will you need to carry heavy loads or equipment? If so, your car should have lots of trunk space. You might even consider a truck.

• What is the climate where you live? If you live in a hot climate, you may want an air conditioner. If you live in a cold climate, you may want four-wheel drive or traction control to make it easier to drive in snow.

Practice and Apply

Answer each question below. Explain your answers.

1. John lives 2 blocks from work. Should he consider a car to get to work?

2. Jen lives 25 miles from work. What should she consider before deciding how to get to work?

3. Elena gets a new job as a photographer for a newspaper. Driving to locations to take pictures for stories is an important part of her work. She finds a car to buy. The car is 10 years old and has over 100,000 miles on it. Should Elena buy this car?

Write the letter of the correct answer.

4. What is a standard transmission?
    A. an automatic transmission
    B. a stick shift
    C. a cold weather feature

5. Which statement is false?
    A. If you live in a cool climate, you may want an air conditioner.
    B. Using public transportation can be a way to save money.
    C. You should think about what you need before you shop for a car.

6. Which feature on your car is worth a discount on your car insurance?
    A. a stick shift
    B. an air conditioner
    C. air bags that reduce your risk of injury in an accident

7. **IN YOUR WORLD** Think about what you learned in this lesson. Which type of transportation is best for you? Explain your choice.

Which Car Is the Best Buy?

How can you find out about the best cars to buy? You can get information in books, magazines, or on the Internet. You can learn from these sources how well certain cars drive, how much gas the cars use, and how much the cars cost. Cars are even rated from the best buy to the worst buy.

A book called a **car price guide** can be very useful. You can find these books in the library and online. The car price guides give you the **invoice price** of a new car. This is the price the new car costs the car dealer. The car price guides also give you the **manufacturer's suggested retail price (MSRP).** This is the price put on the window sticker for the customer.

Car price guides also help you find the price you can expect to pay for a used car. They tell you the suggested retail price for a used car. This price depends on the equipment in the car, its condition, and the number of miles it has been driven. If the car dealer is asking more than the suggested retail price, then the car is not a very good buy.

▶ **EXAMPLE**

Ricky found a used car in excellent condition at a car dealer. The car does not have any special equipment. It was driven a total of 20,000 miles. The dealer wants $7,000 for the car. The price guide says the suggested retail price is $4,500. What is the difference in the prices? Should Ricky buy the car at that dealer?

STEP 1 Compare the prices. $7,000 > $4,500

STEP 2 Subtract to find the difference. $7,000 − $4,500 = $2,500

The difference in the prices is $2,500. Ricky should not buy the car at that car dealer because the dealer's price is too high.

Practice and Apply

Solve each problem. Use the listing for new cars from the car price guide below.

Car	Invoice Price	MSRP
Car A	$6,000	$9,500
Car B	$8,500	$10,900
Car C	$12,800	$13,600
Car D	$13,800	$14,700

1. What is the MSRP for Car A?

2. What is the difference between the invoice price and the MSRP for Car B?

3. What is the difference between the invoice price and the MSRP for Car C?

4. What is the difference between the MSRP of Car C and Car D?

Solve each problem. Show your work.

5. Janet finds a used car at a dealer for $5,800. The suggested retail price in the car price guide is $5,200. What is the difference in the prices? Should Janet buy the car at that dealer?

6. William is looking at a car with a suggested retail price of $4,900. The dealer is asking $4,795. Is this a good deal? Explain.

7. **IN YOUR WORLD** Do you live in a state that has a state sales tax? How much tax, if any, would you need to pay for a car that cost $8,500? (Hint: The Internet can help you find out if your state has a sales tax. Go to www.taxadmin.org)

What About Shopping for a Reliable Car?

You have studied different cars and compared prices. Now you should look at the cars themselves. You may choose to buy a new car or a used car.

There are many ways to shop for a used car. You can look in the classified section of a newspaper. A used car magazine also lists cars for sale. You can even shop for a used car on the Internet. Some people buy used cars from new-car dealers.

Here are some tips for finding a car dealer:

- Ask friends, relatives, or co-workers for the name of a car dealer.

- Go to a car dealer who is a member of the National Automobile Dealers Association (NADA).

- Buy from a dealer who answers all of your questions. A good dealer will not hurry you to make a decision.

Now you think you have found the right car for you. Here's how to be sure you are making a good decision:

- Compare the car dealer's price to the suggested retail price in a car price guide.

- Get other car dealers' prices for the same type of car. Be sure the total miles driven, the equipment, and the condition of each car is about the same. Then compare prices.

- Ask about a **warranty.** A warranty covers major repairs, such as replacing an engine or radiator, during a certain period of time or up to a certain mileage. Car dealers may offer an extended warranty for a fee. This covers repairs for a longer period of time or for more miles.

- Test drive the car. Bring a mechanic with you or have a mechanic check the car for any problems.

Here's a Tip!
You can make an *offer* on a car you like. Check the car price guide. Then make an offer to buy the car for a price that is more than the invoice price but less than the retail price.

Wordwise:
The *condition* of a car describes how the car looks and how well it runs.

Practice and Apply

Use the chart below to solve each problem.

Retail Prices for a Sedan	
Car Dealer	**Retail Price**
Cars R Us	$4,050
We Sell Cars	$3,900
Lots of Cars	$5,200

1. Which dealer has the highest retail price? The lowest retail price?

2. A car price guide lists $4,100 as the suggested retail price for the car. Which car dealer's price is higher than the suggested retail price from the car price guide? How much higher?

3. Which car dealer has the best price for this car? Explain your answer.

4. Cars R Us offers an extended warranty for 1 year or until the car has been driven 50,000 miles, whichever comes first. When you buy the car, it already has been driven 39,670 miles. How many miles can you drive it in the next 12 months before the warranty expires?

5. **WRITE ABOUT IT** Prices for the same car might be very different from one car dealer to the next. Write a short paragraph explaining how you would shop for the best price.

Maintaining Skills

Find each percent.

1. 12% of $100

2. 18% of $4,000

3. 25% of $2,000

4. 7% of $1,400

5. 60% of $80

6. 5% of $700

You have found the car you want to buy. How will you pay for it? Here are some choices:

- Use your savings and pay cash for the car.

- Take out a car loan from the car dealer or a bank and pay it back in monthly payments.

If you take out a loan, you need to decide how much you can afford for a **down payment**. A down payment is an amount of cash you pay to the car dealer. The down payment is subtracted from the total cost of the car. The greater the down payment, the less money you need to borrow to buy the car.

Different lending institutions offer auto loans. Find the one that offers the lowest interest rate for the monthly payments you can afford.

Wordwise

Gap insurance covers the remaining loan payments if a car is stolen or destroyed in an accident before the loan is paid off.

▶ **EXAMPLE 1**

You want to buy a car that has a total cost of $4,380. You will pay 15% of the car's total cost as a down payment. How much do you need to borrow to buy the car?

STEP 1 Change the percent to a decimal.

15% = 0.15

STEP 2 Multiply the car's total cost by the decimal to find the amount of the down payment.

$4,380 × 0.15 = $657

STEP 3 Subtract the down payment from the car's total cost.

$4,380 − $657 = $3,723

Consumer Beware!

Other charges can be added to the retail price of the car. In some states, you must pay sales tax on the price of the car. Other costs might include transportation from the factory and a fee for license plates.

You need to borrow $3,723 to buy the car.

If you already have a car, you can use it as a **trade-in**. You sell the car dealer your old car and use the money to pay part of the cost of the new car.

► **EXAMPLE 2**

Tawanda wants to buy a newer car. She will use her old car as a trade-in. The car she will buy has a total cost of $4,400. The car dealer offers Tawanda $2,500 for her trade-in. Tawanda will take out a loan to buy the car. How much does Tawanda need to borrow?

Did You Know?
If you trade in your old car and also give a down payment, you can lower the amount of your loan.

Subtract the trade-in amount from the cost of the newer car.

$4,400 − $2,500 = $1,900

Tawanda needs to borrow $1,900.

Practice and Apply

Complete the chart. Use a calculator if you like.

	Cost of Car	Percent of Down Payment	Amount of Down Payment	Amount of Loan
1.	$3,500	10%	?	?
2.	$4,300	15%	?	?
3.	$5,600	20%	?	?
4.	$5,900	15%	?	?
5.	$7,400	20%	?	?
6.	$8,200	10%	?	?

Solve each problem. Show your work.

7. Elwood is buying a newer car that costs $5,200. He trades in his old car for $2,600. He also makes a $500 down payment. How much does Elwood need to borrow to buy the car?

8. CRITICAL THINKING Gregg wants to buy a car. The car costs $6,000. He pays $1,500 as a down payment. What percent of the total cost is the down payment?

How Do You Check the Contract?

When you take out a car loan, you sign a contract, or agreement. If there is anything in this contract you do not understand, ask the banker or the car dealer. If you still have questions, ask your city's legal aid office.

Read the contract carefully to make sure it is correct. Here are some questions to ask yourself before signing:

- Are all the blanks in the contract filled in?
- Has the correct amount of the down payment been subtracted from the car's total cost?
- How many months is the loan?
- How much are the monthly payments? When are they due?
- What happens if a payment is late?

David is buying a used car. Here is a part of his contract.

Did You Know?

Each state has *lemon laws* to protect consumers who buy defective vehicles. Each state's lemon laws are different. Check your state's lemon laws at the library or on the Internet.

BUYER: David R. Simms
          12 Willow Ave., Roselle, CT 08945
	SELLER: Cars R Us
	          146 Rt. 49 East, Fairlawn, CT 08976

NEW OR USED CAR: Used	MODEL YEAR: 2000
MAKE AND MODEL: 4-door Chevel	VEHICLE IDENTIFICATION NUMBER: 1G878R345X6747663
ANNUAL PERCENTAGE RATE: 7%	NUMBER OF PAYMENTS: 24
PAYMENTS DUE MONTHLY STARTING: 10/05/03	AMOUNT OF PAYMENTS: $129.39

LATE PAYMENTS: A late charge of 10% of the payment is required if a payment is received 5 days after the due date.	
CASH PRICE OF CAR	+ $4,000
AMOUNT OF SALES TAX (SALES TAX RATE = 6%):	+ $240
TOTAL OF OTHER CHARGES:	+ $150
AMOUNT OF DOWN PAYMENT:	− $1,500
UNPAID BALANCE: (AMOUNT OF LOAN)	= $2,890

Practice and Apply

Use David's car contract on page 304 to answer each question.

1. Which car dealer is selling a car to David?

2. Is David buying a new or used car?

3. Is David's car a 2-door or a 4-door?

4. How long is the term of David's auto loan in years?

5. When is David's first monthly payment due?

6. How much will the total monthly payment be if it is 5 or more days late? (Hint: Find 10% of the amount of the monthly payment. Then, add this to the amount of the monthly payment.)

7. What is the total purchase price of the car? (Hint: Add all the costs listed on the contract above the down payment.)

8. How much money did David give as the down payment?

9. How much money will David borrow to buy the car?

10. **IN YOUR WORLD** You read your contract but cannot find information on late payments. What do you do?

Maintaining Skills

Compute.

1. $\begin{array}{r} \$5,398.00 \\ -\ 3,489.40 \end{array}$	2. $\begin{array}{r} \$3,675 \\ \times\quad 0.08 \end{array}$	3. $\begin{array}{r} \$4,876.95 \\ -\quad 341.39 \end{array}$

4. $6,496 ÷ 4

5. $398 × 24

6. $315.25 × 36

Find each percent.

7. 10% of $691

8. 40% of $300

9. 20% of $850

What About Making Car Loan Payments?

You make a down payment of $600 on a $1,600 car. You still need $1,000 more to buy the car. You decide to get a car loan from your bank.

The chart below shows the amount of monthly car payments with different interest rates and terms.

Monthly Payments for a $1,000 Loan				
Interest Rate	12 Months	24 Months	36 Months	60 Months
6%	$86.07	$44.32	$30.42	$19.33
7%	$86.53	$44.77	$30.88	$19.80
8%	$86.99	$45.23	$31.34	$20.28
9%	$87.45	$45.68	$31.80	$20.76
10%	$87.92	$46.15	$32.27	$21.25
11%	$88.39	$46.61	$32.74	$21.74
12%	$88.85	$47.08	$33.22	$22.24

▶ **EXAMPLE 1**

Suppose you get a car loan for $1,000 with an interest rate of 9%. The term of the loan is two years. What is the total amount you will pay on the loan?

STEP 1 To find the monthly payment, first find the row with the 9% interest rate.

STEP 2 Now find the column for the term of the loan. Change 2 years to months.
$2 \times 12 = 24$

STEP 3 Find where the row and column meet. $45.68
This is the monthly payment.

STEP 4 Multiply the amount of the monthly payment by the number of months.
$45.68 \times 24 = \$1,096.32$

The total amount you will pay is $1,096.32.

EXAMPLE 2

Now you can find the total interest you will pay.

You will pay $1,096.32 altogether for the loan. So, what is the total interest you will pay?

Subtract the amount you borrowed.
$1,096.32 − $1,000.00 = $96.32

The total interest you will pay is $96.32.

Practice and Apply

Use the chart on page 306. Find each monthly payment and total amount paid.

	Amount of Loan	Interest Rate	Term	Monthly Payment	Total Amount Paid
1.	$1,000	8%	1 year	?	?
2.	$1,000	12%	3 years	?	?
3.	$1,000	7%	5 years	?	?
4.	$1,000	9%	1 year	?	?
5.	$1,000	6%	2 years	?	?
6.	$1,000	11%	3 years	?	?

Solve each problem. Show your work. Use the chart on page 306.

7. Susan has a $1,000 car loan. The loan has an interest rate of 11% and a term of 12 months. How much interest will she pay on the loan altogether?

8. Nora has a $1,000 car loan. The loan has an interest rate of 6% and a term of 36 months. How much interest will she pay on the loan altogether?

9. CRITICAL THINKING Ray can afford to pay $75 as a monthly payment for his car loan. Can he afford a $1,000 car loan with an interest rate of 7% and a term of 36 months? If so, how much would he have left over from his $75?

What About Leasing a Vehicle?

Instead of buying a car, you decide to **lease** one. You sign a contract to lease the car for a period of time.

Leasing a car is different from buying. The monthly payments may be lower than a car loan. You also do not need to make a down payment.

Here's a Tip!
Gap insurance is also a good idea for a lease.

When the term of the lease is over, you do not own the car. You have two choices. You may return the car to the car dealer or you may buy it. If you choose to buy this car, you pay the **residual value** of the car. The residual value is the retail price of the car minus the amount of its **depreciation.**

▶ **EXAMPLE 1**

At the end of the lease, the value of the car has depreciated by 40%. The retail price of the car is $12,000. How much is the residual value?

STEP 1   Change the percent to a decimal.
40% = 0.4

STEP 2   Multiply the retail price by the decimal to find the amount of depreciation.
$12,000 × 0.4 = $4,800

STEP 3   Subtract the amount of depreciation from the retail price.
$12,000 − $4,800 = $7,200

The residual value of the car is $7,200.

Wordwise
Agreement is another word for *contract.*

So if you want to buy this car at the end of the lease, the retail price is $7,200. This retail price would be written into the lease agreement.

The lease agreement also includes rules. You should read it carefully before signing. One rule limits the total number of miles you are allowed to drive the car during the term of the lease. If you drive more than this number of miles, you are charged a fee for each mile over the limit.

▶ **EXAMPLE 2**

You are leasing a car. Your lease limits your mileage to a total of 36,000 miles. When your lease ends, the car has been driven 37,950 miles. The fee is $0.15 for each extra mile. How much do you owe for the extra miles?

STEP 1 Subtract the number of miles allowed in the lease from the number of miles driven.
37,950 − 36,000 = 1,950 miles

STEP 2 Multiply the number of miles over the limit by the fee.
1,950 × $0.15 = $292.50

You owe $292.50 for the extra miles.

Practice and Apply

Complete the chart.

	Retail Price of Car	Percent of Depreciation	Amount of Depreciation	Residual Value of Car
1.	$8,500	35%	?	?
2.	$9,200	40%	?	?
3.	$7,800	45%	?	?

Solve each problem. Show your work.

4. Priscilla's lease limits her mileage to 36,000 miles. When her lease ends, the car was driven 38,211 miles. The fee is $0.12 for each extra mile. How much does she owe for the extra miles?

5. **CRITICAL THINKING** Erin leases her car. A $213 payment is due on the 18th of the month. The late fee is $35 if the payment is not received by that day. Erin forgets to send the August check until August 16. What should she do? What amount should she write on the check if she sends it regular mail?

15·8 ▶ Problem Solving

Solve each problem. Show your work.

1. Al finds a car he likes at Car World. The price is $5,500. In the car price guide, the suggested retail price for this same car is $5,100. What is the difference in price? Should Al buy the car at Car World?

2. The total cost of the car John wants to buy is $4,500. He makes a down payment of $2,000. How much does John need to borrow?

3. Rob is buying a car for $11,500. He pays a 6% sales tax and then pays $1,350 in additional charges. He makes a $1,500 down payment. How much does Rob need to borrow to buy the car?

4. **OPEN ENDED** The invoice price on a new car is $11,999. The manufacturer's suggested retail price is $12,994. What should a buyer expect to pay for the car? If she puts $2,000 down, how much will she need to borrow?

Calculator

Your monthly car payments are $307 for 24 months. You made a $2,000 down payment. You can use a calculator to find the total amount you pay.

Press: 3 0 7 × 2 4 + 2 0 0 0 = 9,368

Find the total amount paid for each car.

	Monthly Payment	Term of Loan	Down Payment	Total Paid
1.	$150	60 months	$2,000	?
2.	$175	48 months	$1,500	?
3.	$200.75	36 months	$2,500	?
4.	$275.90	24 months	$3,000	?

DECISION MAKING:
Should I Buy a New or Used Car?

Martha is 21 years old. She works and goes to college part time. She found a used car for $4,000 and a new car for $10,000. She is having trouble deciding which car to buy. She did some research on the Internet. She will also ask her friend, a mechanic, to look at the used car before she buys it. Martha made a list of advantages and disadvantages of buying a used car.

Buying a Used Car	
Advantages	**Disadvantages**
• Lower retail price • Lower monthly payment, if any	• Limited warranty, if any • Limited choice of equipment and colors • At least 2% higher interest rates for an auto loan for a used car • Hard to get an auto loan for a car more than 4 or 5 years old • Lower trade-in value, if any

1. Martha has saved $2,500 for a down payment. She wants her down payment to cover at least 25% of the cost of the car. What is the most that she can pay for a car?

2. Martha will need to get an oil change on the new car every 6,000 miles. Martha will need to get an oil change on the old car every 3,000 miles. Martha drives about 12,000 miles per year. An oil change costs $30. About how much will Martha spend in 1 year on oil changes if she buys the new car? The used car?

You Decide

Martha will use $2,500 as a down payment. The new car costs $10,000 and has a 36-month or 36,000-mile warranty. The used car costs $4,000 and has a 90-day or 3,000-mile warranty. Martha can get a car loan at 5% interest on the new car or a car loan at 8% interest on the used car. Which car do you think Martha should buy? Explain.

Summary

The best vehicle for you to buy depends on what you do and where you live. It also depends on how much you can afford to pay each month for car payments, car insurance, maintenance, repairs, and gas.

The information in car price guides can help you compare car dealers' prices to see if they are fair.

You can take out a loan to pay for a new or used car. The greater your down payment, the less money you need to borrow.

If you take out a car loan, check that all the information in your loan contract is filled in correctly, based on the information you know.

Monthly loan payments depend on the amount borrowed, the interest rate, and the time you take to repay the loan.

You can lease a car instead of buying one. You will not own the car when the lease ends.

The lease agreement will tell you information such as the residual value of the car and the fees you may have to pay when the term is over.

depreciation

down payment

invoice price

manufacturer's suggested retail price

residual value

trade-in

Vocabulary Review

Complete the sentences with words from the box.

1. The value of a car at the end of a lease is the _____.

2. The money you give a car dealer that lowers the amount left to pay on a car is a _____.

3. A _____ is a used vehicle that a car dealer takes from a customer in part payment for another vehicle.

4. The _____ is the price suggested by the manufacturer for a customer to pay for a new car.

5. A decrease in the value of an item because of use and age is _____.

6. The _____ is the price the car dealer pays the manufacturer for a car.

Chapter Quiz

Solve each problem. Show your work.

1. Andrea has found a car that she wants. The manufacturer's suggested retail price is $12,000. The dealer is asking $11,500 for this car? Is this a good deal? Why or why not?

2. Tyrone will pay 25% of the total cost of a car as a down payment. The car's total cost is $6,000. How much will his down payment be?

3. Bill is buying a car that costs $6,500. He gave the car dealer $2,500 in cash as a down payment. He also used his old car as a trade-in. The dealer gave him $1,500 for his old car. How much does Bill need to borrow?

4. Your 2-year car loan has a monthly payment of $180.92. What is the total amount you will pay on this loan?

5. Alexis is leasing a car. By the end of the lease, the value of the car has depreciated by 45%. The retail price of the car was $8,500. How much is the residual value of the car?

Maintaining Skills

Subtract.
1. $3,500 − $1,250

2. $8,000 − $1,500

3. 38,120 − 36,000

4. 31,010 − 24,000

Find each percent.
5. 8% of 30

6. 30% of 70

7. 2% of 700

8. 12% of $560

9. 80% of $400

10. 70% of $920

Chapter 16

Maintaining a Vehicle

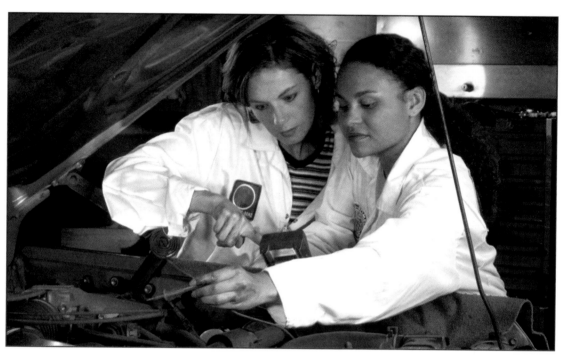

Many people enjoy working on their cars. Why is it a good idea to know something about how a car works?

Learning Objectives

LIFE SKILLS

- Know what information you need to give to an insurance agent to get car insurance.
- Understand the different types of insurance coverage you can get for your car.
- Understand why the cost of car insurance may vary.
- Understand how a deductible works.
- Understand how to take care of your car.
- Analyze a repair bill.

MATH SKILLS

- Add and subtract whole numbers.
- Add and subtract with money.
- Find an average.

Words to Know

coverage	protection against the costs of accidents, theft, or other risks
liability coverage	insurance that pays for other people's injuries and property damage if you cause an accident
uninsured motorist coverage	insurance that pays for some of your costs if a driver without insurance hits your car
collision coverage	insurance that pays for the repair of your car after an accident
comprehensive coverage	insurance that covers damage to your car from things such as fire and flood
medical coverage	insurance that covers your medical bills after an accident
premium	money paid to a company to buy insurance
deductible	the amount of money you must pay for repairs or injuries before your insurance company pays
odometer	an instrument that records how many miles a car has been driven
labor	work done on a car in a repair shop

Project: Buying Insurance for Your Car

Visit the federal consumer Web site
www.pueblo.gsa.gov/cic_text/cars/autoinsu/autoinsu.htm to
learn about auto insurance. This Web site offers hints on
how to lower your insurance costs. Choose a car advertised
in the local paper that you would like to own. Then contact
three insurance companies to find out what a student with
a good driving record would pay to insure the car. Make a
chart or graph to organize the information you gather.
Which company offers the best rate?

What About Car Insurance?

Once you get a car, you may want to drive it right away. Before you do, you must insure your car. Car insurance helps to pay the costs after a car accident. It also helps to pay for any damage that happens to your car when you are not in it.

Your insurance plan should give you **coverage** for both personal injury and property damage. The chart below shows the different types of coverage you can buy.

Insurance Coverage	
Liability coverage	Pays for other people's personal injury and property damage if you cause an accident
Uninsured motorist coverage	Pays for some of your costs when a driver who is not insured causes an accident
Collision coverage	Pays to repair your car after an accident
Comprehensive coverage	Pays for damage that is caused to your car by things such as fire and flood
Medical coverage	Pays your medical bills if you are injured

When you buy insurance, the insurance agent will ask you several questions. The **premium**, or price, that you pay depends on your answers.

Here's a Tip!
Your insurance company will send you an insurance card with your policy number. Always keep the card in a safe place inside your car.

- Are you male or female?

- How old are you?

- Does your driving record include any accidents or speeding tickets in the past few years?

- What kind of car do you drive?

- How far do you drive to work or school?

- Are you married or single?

- Where do you live?

Each state has laws about the insurance that drivers must have. Almost all states require liability coverage, and many require uninsured motorist coverage as well.

How can you find the best buy in auto insurance? Ask people you trust for the name of a good insurance company. Talk to agents at different companies to find out about rates, or prices. Compare these rates.

Practice and Apply

Write the letter of the correct answer.

1. Which type of insurance covers your car in case of a fire?

   A. collision
   B. liability
   C. comprehensive

2. Which type of insurance pays for damage to another car if you cause an accident?

   A. collision
   B. liability
   C. comprehensive

3. What does collision insurance cover?

   A. the repair of your car
   B. medical expenses
   C. the repair of a mailbox you damage with your car

4. Which information must you give the insurance agent?

   A. the kind of car that you drive
   B. the name of your best friend
   C. the salary that you earn

5. **IN YOUR WORLD** Find out what car insurance coverage is required in your state. Use the Internet or go to the library. Write about what you learn.

You have choices to make when you buy car insurance. The more coverage you buy, the higher your premium. More coverage also means that you may pay less for repairs and other expenses if you have an accident.

The cost of insurance depends on many things. If you live in a rural area, you will pay less than if you live in a city. Insurance rates are higher if you are single, male, under 25, or drive an expensive sports car.

Did You Know?
New Jersey, Washington, D.C., New York, Massachusetts, and Rhode Island have the highest auto insurance rates in the country.

Some insurance companies offer a discount if you have a good driving record. Another discount may be available if you are a good student. You may get a discount if you complete a driver training class. Special rates are available if you insure more than one car or your car has low annual mileage.

Insurance rates differ, depending on where you live. The chart below gives an example of this.

Liability Coverage 15/30/5	
Region	Basic Annual Rate
A	$615
B	$420
C	$258
D	$460

Here's a Tip!
The liability coverage numbers 15/30/5 tell you the amount of coverage you have. Each number represents thousands of dollars.

The chart shows typical rates for the following liability coverage limits:

15 ⟶ $15,000 maximum payment for each person injured

30 ⟶ $30,000 maximum payment for all personal injuries

5 ⟶ $5,000 maximum payment for property damage

Practice and Apply

Use the chart on page 318 to answer each question.

1. Which region has the highest rate? What type of area do you think this might be?

2. Which region has the lowest rate? What type of area do you think this might be?

3. What is the difference between the highest liability rate and the lowest liability rate?

4. Tom has 30/35/10 liability coverage on his car. What is the maximum the insurance company will pay for each person he injures in an accident?

5. Which policy offers a higher maximum payment for property damage, a policy with 10/10/5 liability coverage or a policy with 15/35/5 liability coverage?

6. **WRITE ABOUT IT** Write a letter to an insurance company, asking about its insurance rates. Give information about yourself, the coverage you want, and the car you own or would like to own.

Maintaining Skills

Compute.

1. $10,375 - 8,246$ 2. 37.95×4 3. $12,376 + 7,500$

4. $12,832 - 6,131$ 5. $1,021 \times 12$ 6. $472.95 + 111.42 + 16.18$

16·3 How Does a Deductible Work?

You can lower the cost of your insurance by increasing the amount of your **deductible**. The deductible is the greatest amount you must pay to fix your car after an accident. The insurance company pays any amount above that if more money is needed to make the repair. You can have a deductible on collision and comprehensive coverage. The higher your deductible, the less your insurance will cost.

▶ **EXAMPLE**

You have a $250 deductible on collision insurance. You hit a car in a parking lot. It costs $586.54 to fix your car. How much do you pay for the repair? How much does the insurance company pay for your car repair?

Subtract the deductible from the total repair cost.

$586.54 − $250.00 = $336.54

You pay $250. The insurance company pays $336.54.

Practice and Apply

Solve each problem. Show your work.

1. Glen has a collision deductible of $200. He scratched his car while backing out of his garage. It costs $285.12 to fix the scratch on his car. How much will Glen pay? How much will the insurance company pay?

2. Sally has a comprehensive deductible of $100. A branch fell on her car and dented the roof. The repair cost $389.70. How much will the insurance company pay?

3. **CRITICAL THINKING** Rick has a collision deductible of $500. He dented the side of his car. It costs $450.20 to repair the car. How much will Rick pay? How much will the insurance company pay? Explain.

16·4 What About Keeping Your Vehicle Running?

If your car has a warranty, the car dealer has promised to make certain repairs at no cost to you. However, you must take care of your car to keep it running well.

Consumer Beware!
A maintenance schedule tells you how to care for your car and when to replace certain parts. The warranty on a new car may not cover necessary repairs if you have not followed the maintenance schedule.

Check the fluid levels in your car often. Fluids include brake fluid, gasoline, oil, water, antifreeze, transmission fluid, and windshield washer fluid. Check the air pressure in your tires. Rotate your tires on schedule. Replace your windshield wiper blades, hoses, and belts when they become worn. Check the battery regularly, especially before cold weather. Use the **odometer** to keep track of your car's mileage.

Every car has an owner's manual. Use it to answer questions such as these:

- What is the maintenance schedule for the car?

- How often should the oil be changed?

- How often should the tires be rotated?

Practice and Apply

Solve each problem. Show your work.

1. The last time you changed your oil, the odometer read 11,620 miles. Now it reads 13,999 miles. How many miles have you driven since the last oil change?

2. You change your oil every 3,000 miles. The last time you had your oil changed, your odometer read 28,220. Now it reads 29,520. How many more miles can you drive before your next oil change?

3. **IN YOUR WORLD** What does an oil change and engine tune-up cost for your car or for a car that you would like to own? (Hint: Use the Internet, ads in the newspaper, or local repair shops to find the information.)

16·5 ▶ What About Car Repairs?

If your car isn't running properly, take it to a mechanic immediately. A small problem may become expensive if you wait. It can also be dangerous.

First get an estimate for the cost to fix your car. This will help you decide how to pay for the repair. The estimate and the actual costs should be almost the same. Otherwise the mechanic should call you before adding other charges to the bill.

After the car is repaired, you will get a bill describing the work that was done and the charges. Be sure you understand everything on the bill before you pay it.

Repair bills may look different, but all should contain the same information. Look at the bill on page 323. At the top it has the owner's name and address. The bill also shows the date of the repair, the mileage on the odometer, and other information about the car.

The left side of this bill shows the new parts that were used to fix the car. The cost of each part is also listed. The work that the mechanic does is called **labor.** The labor is listed under DESCRIPTION OF WORK. The cost of the labor is listed next to the description.

At the bottom of the bill it says AUTHORIZED BY. Before the repairs are done, the customer signs here to give the mechanic permission to work on the car.

Practice and Apply

Use the repair bill on page 323 to answer each question.

1. What kind of car is being repaired?

2. What was the mileage on the car when it came in for repair?

3. Who authorized the work to be done?

4. What is the total cost of the labor?

5. Tax is only paid on the parts. How much tax will be paid? Round to the nearest cent.

6. What is the total cost of the repair bill?

7. **CRITICAL THINKING** Arnie paid cash for an $859 car repair. He paid with 7 one-hundred-dollar bills, 2 fifty-dollar bills, 2 twenty-dollar bills, and 2 ten-dollar bills. How much change should he receive?

QTY.	PART	PRICE					
Name: Sharon Goss							
Address: 487 Hawthorne Drive			**City:** Temecula				

Name: Sharon Goss

Address: 487 Hawthorne Drive   **City:** Temecula

QTY.	PART	PRICE	DATE 6/14/03		ORDER NO.	WHEN PROMISED 6/16/03	PHONE 555-1125
1	Oil Filter	3	95	**YEAR & MAKE OF CAR–TYPE OR MODEL** 1999 Custom Cruiser		**ESTIMATE** $945	**SERIAL NO.**
2	Front Disk Pads	79	95				**MOTOR NO.**
2	Rotors	284	40	**LICENSE NO.** 245-MWW		**MILEAGE** 67,458	**WRITTEN BY** Sam
1	Muffler	75	80				
1	Gasket	11	20	**DESCRIPTION OF WORK**		**AMOUNT**	
6	Spark Plugs	36	50	Turn rotors and repack bearings		69	95
				Fix exhaust leak		47	50
				Major tune-up		185	00

	TOTAL PARTS	575	70	GAS, OIL & GREASE	CHECK BELOW	LABOR ONLY	?	?
ESTIMATES ARE FOR LABOR ONLY, MATERIAL ADDITIONAL				LBS. GR.	LUBRICATE	PARTS	?	?
				GALS. GAS	ENGINE OIL			
				QTS. OIL 5 of 10/30 12 50	TRANS.	GAS, OIL & GREASE 12	50	
I HEREBY AUTHORIZE THE ABOVE REPAIR WORK TO BE DONE ALONG WITH NECESSARY MATERIALS.				TOTAL GAS, OIL & GREASE 12.50	TOTAL SERVICE	TAX 6%	?	?
				AUTHORIZED BY *Sharon Goss*		TOTAL	?	?

Solve each problem. Show your work. Use a calculator if you like.

1. Al sends his insurance company 3 repair bills for three different collisions. The bills are for $550, $400, and $650. The policy has a $200 deductible on each claim. How much will the company pay?

2. Laura drove 94 miles on Wednesday, 123 miles on Thursday, and 53 miles on Friday. What was the average number of miles she traveled per day?

3. Kelly sells car insurance. She just sold a policy with the following annual premiums: liability—$512.48; medical—$110; uninsured motorist—$72; collision—$398.42; comprehensive—$153.50. What is the total annual premium for this policy?

4. **OPEN ENDED** An owner's manual recommends an oil change every 3 months or every 7,500 miles if you drive mostly on the highway. The schedule is every 3 months or every 5,000 miles if you drive mostly on local roads. Describe a driver whose driving needs require changing the oil every 2 months.

 Calculator

Gas costs $1.379 per gallon. You can use a calculator to find out how many gallons you can buy for $10. Round down to the nearest tenth.

Press: ⟦1⟧⟦0⟧⟦÷⟧⟦1⟧⟦.⟧⟦3⟧⟦7⟧⟦9⟧⟦=⟧ ⟦ 7.25 163 16 ⟧

You can buy about 7.2 gallons of gas.

Find the number of gallons that $20 will buy at each price. Round down to the nearest tenth.

1. $1.459 per gal   2. $1.579 per gal   3. $1.849 per gal

ON-THE-JOB MATH:
Auto Mechanic

Bruce is an auto mechanic for Cars-R-Us. He is 23 years old. He started working for the company in high school. Then he went to a trade school for mechanics. He has been with Cars-R-Us for several years.

Bruce specializes in maintenance and repairs on foreign cars. He inspects all the fluid levels, belts, and hoses in the car. If anything needs to be replaced, he lets the customer know.

He prepares bills for the cost of his labor and the parts he uses. He also puts a sticker on the car that shows the mileage and date for the next oil change. This reminds the customer when to return for the next oil change.

Bruce buys his own tools. He learned to recognize quickly which size wrench he needs for a particular repair. He also knows how to use diagrams and repair manuals for the different cars.

Solve each problem.

1. Bruce just changed the oil in a car. The odometer reads 28,915. The car needs another oil change in 5,000 miles. What mileage should he write on the sticker?

2. Bruce spent 4 hours working on a car. The shop charges $65 per hour. How much should he charge for labor?

3. The shop keeps oil in 55-gallon drums. Bruce used 38 gallons of oil from a full drum this week. How much oil is left in the drum?

You Decide
Bruce earns $8 per hour. He also makes a 20% commission on all his labor charges. His boss has offered to pay him a fixed weekly salary instead of an hourly rate plus commission. The new salary would equal his average weekly pay for last year. Would you accept the new salary offer if you were Bruce? Explain.

Summary

You must give some basic information about yourself and your car to the insurance agent when you apply for car insurance.
Car insurance gives you coverage for personal injury and property damage.
The cost of car insurance depends on many things, including your age, your driving record, and where you live and work.
Subtract your deductible from the total cost of a repair to find how much the insurance company will pay.
Follow the maintenance schedule in your owner's manual to keep your car in good running condition.
A repair bill tells you the cost of a repair. It lists the parts and labor, and the cost of each. Read it carefully before you pay the bill.

collision coverage

comprehensive
   coverage

deductible

labor

liability coverage

odometer

premium

Vocabulary Review

Complete the sentences with words from the box.

1. ____ pays for other people's injuries and property damage if you cause an accident.

2. ____ pays for the repair of your car after an accident.

3. A ____ is money paid to a company to buy insurance.

4. A ____ is the amount of money you must pay for repairs or injuries before your insurance company pays.

5. ____ pays for damage to your car from things such as fire and flood.

6. Work done on a car in a repair shop is called ____.

7. An instrument that records how many miles a car has been driven is an ____.

Chapter Quiz

Solve each problem. Show your work.

1. What type of car insurance covers you in case you are injured while driving your car?

2. One policy has 10/10/10 liability coverage. Another policy has 15/30/5 liability coverage. Which policy offers a higher maximum payment for all personal injury claims in an accident?

3. Suppose you park your car in a parking lot. Someone damages the door of your car while you are gone. The repair costs $497. Your deductible is $250. How much would you expect the insurance company to pay?

4. You need to change your wiper blades every 6,000 miles or every 6 months, whichever comes first. The last time you changed them was in March, when the odometer read 4,550 miles. When should you change them again?

5. Rick took his car to the repair shop for an oil change and a tune-up. The parts and tax cost $26.50. The total bill was $209.73. How much did Rick pay for labor?

Maintaining Skills

Estimate each sum.

1.	$615.37	**2.**	$404.51
	+ 289.34		+ 396.65

3. $101.32
+ 79.65

4.	$30.25	**5.**	$704.58
	49.25		199.36
	+ 51.23		+ 102.13

6. $11.29
9.67
+ 21.32

Unit 8 **Review**

Write the letter of the correct answer. Use the graph to
answer questions 1 and 2.

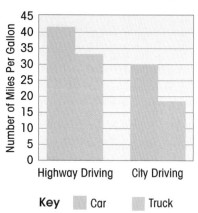

Miles Traveled on a Gallon of Gas

Key   Car   Truck

1. How many more miles per gallon
   of gas does the car get than the
   truck when traveling on a
   highway?

   A. 3 miles per gallon
   B. 9 miles per gallon
   C. 12 miles per gallon
   D. 15 miles per gallon

2. How many miles can the car drive
   in the city on two gallons of gas?

   A. 36 miles
   B. 30 miles
   C. 60 miles
   D. 18 miles

3. Jacob is buying a car for $14,500.
   He gives a 20% down payment.
   How much will he borrow?

   A. $2,900
   B. $11,600
   C. $10,875
   D. $3,625

4. The retail price of a leased car
   is $9,500. At the end of the
   lease, it had depreciated by 50%.
   How much is its residual value?

   A. $9,500
   B. $19,000
   C. $950
   D. $4,750

5. Lou's car window was broken by a
   rock. It costs $642 to replace. Lou's
   comprehensive deductible is $250.
   How much will the insurance
   company pay to fix it?

   A. $524
   B. $642
   C. $392
   D. $742

Challenge

Amy is buying a car for $8,000. She
gives a $2,500 down payment. She
borrows the rest of the money at
6.5% simple interest to be paid
back over 3 years. About how much
does Amy pay each month?

Unit 9 — Recreation, Travel, and Entertainment

Chapter 17 **Budgeting for Recreation**

Chapter 18 **Planning a Trip**

The distance a car travels depends on its speed and the number of hours the car is on the road.

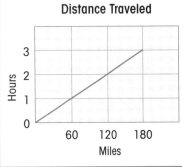

The line graph shows the number of miles a car travels at 60 miles per hour. Use the graph to answer each question.

1. How many miles does the car travel in 1 hour?

2. How many miles does the car travel in 2 hours?

3. How many miles does the car travel in 30 minutes? (Hint: 30 minutes is a half-hour.)

Be sure that you budget for recreation. Look for bargain days and discounts. What are some of the recreational activities you enjoy?

Learning Objectives

LIFE SKILLS

- Find the costs of recreational activities, including equipment and clothing.
- Determine the amount of money saved from a bargain.
- Compare costs for the same activity.
- Determine how to budget for recreation.

MATH SKILLS

- Add, subtract, multiply, and divide with money.
- Find the percent of a number.

Words to Know

spectator sport	a sport or athletic event that people enjoy watching
recreation	activities you like to do to relax in your free time
inventory	a list of supplies, materials, or equipment

Project: Budgeting for Recreation

Choose two or three activities that you would like to do in your free time. Compare the activities by answering the following questions. Make a chart to help you organize your information.

- How many times a month will you do the activity?
- Will you do the activity all year?
- What equipment and clothing will you need?
- How much will it cost to do the activity each time?

Look on the Internet or in newspapers for the information you need. Write a paragraph to tell which activity you would choose first and why.

What do you do with your free time? Perhaps you like to play sports such as tennis, soccer, or volleyball. Maybe you like a **spectator sport** such as professional baseball or basketball. Or you might choose to do an activity that you can do by yourself, such as playing the guitar or reading.

Whatever form of **recreation** you choose, you will need to find out how much it will cost. Some activities will require that you buy or rent equipment or clothing.

▶ **EXAMPLE**

Josh is taking his brother Lucas bowling. Josh is a 19-year-old student. Lucas is 15 years old. How much will it cost altogether for Lucas and Josh to rent shoes and bowl 4 games?

Valley View Bowling Prices are per person per game.	
Adults	$4.75
Seniors (over 65)	$2.75
Students (18–23)	$3.50
Juniors (under 18)	$2.25
Shoe rental	$2.00

STEP 1 Multiply to find the cost of 4 games for Josh.

$3.50 × 4 = $14.00

STEP 2 Multiply to find the cost of 4 games for Lucas.

$2.25 × 4 = $9.00

STEP 3 Multiply to find the cost of the shoes.

$2.00 × 2 = $4.00

STEP 4 Add to find the total cost.

$14 + $9 + $4 = $27

It will cost $27 altogether.

Practice and Apply

Solve each problem. Use the chart on page 332 for Problems 1 and 2. Use a calculator if you like.

1. You go bowling. You are a 20-year-old student. You play 2 games and rent shoes. What is your total cost?

2. Gary and 3 friends go bowling. They are all 15 years old. They each bowl 3 games. Gary pays for the games. Each person pays for his own shoes. How much money does Gary spend?

3. You decide to take guitar lessons. Each half-hour lesson costs $20. How much will you budget to pay for a half-hour lesson each week for 26 weeks?

4. Your sister plans to take weekly tap dancing lessons. Each lesson costs $15. She plans to take 12 lessons. Tap shoes cost $48. There is 5% sales tax on the tap shoes. How much should she budget for the lessons and shoes?

5. Morgan and her brother are going to a baseball game. Each ticket costs $20.00. Hot dogs cost $3.50 each and soft drinks cost $3.00 each. There is 4% sales tax on each item. How much will Morgan spend if she buys 2 of everything?

6. Amir and his sister are going to a bicycle race. Each ticket costs $15.00. Hot dogs cost $3 each and soft drinks cost $2.50 each. Amir buys 2 of each item. There is 5% sales tax on each item. If Amir budgets $50, will he have any money left over? If so, how much?

7. **IN YOUR WORLD** Call a bowling alley in your area. Find out the cost to bowl one game and to rent shoes.

17·2 How Can Bargains Save You Money?

Most businesses have times when they have fewer customers than usual. Some businesses charge lower prices at these times. This could mean a bargain for you.

Sometimes movie theaters have fewer customers during the day than in the evening. Movie ticket prices may be lower during these times. Restaurants, amusement parks, and bowling alleys may also offer bargain days or nights.

► **EXAMPLE**

Josh is a 19-year-old student. He is planning to bowl 3 games on Friday. If he waits until Monday to bowl, how much will he save?

Valley View Bowling Prices are per person per game.	Tuesday– Sunday	Monday is Bargain Day!
Adults	$4.75	$3.50
Seniors (over 65)	$2.75	$1.25
Students (18–23)	$3.50	$2.50
Juniors (under 18)	$2.25	$1.75
Shoe rental	$2.00	$1.75

STEP 1 Multiply to find the cost of 3 games on Friday.

$$\begin{array}{r} \$3.50 \\ \times \quad 3 \\ \hline \$10.50 \end{array}$$

STEP 2 Multiply to find the cost of three games on Monday.

$$\begin{array}{r} \$2.50 \\ \times \quad 3 \\ \hline \$7.50 \end{array}$$

STEP 3 Subtract to find how much Josh could save.

$$\begin{array}{r} \$10.50 \\ - \quad 7.50 \\ \hline \$3.00 \end{array}$$

Josh will save $3.00 if he bowls on Monday.

Practice and Apply

Use the chart on page 334 to solve Problems 1–5.

1. Lucas is 15 years old. He plans to take his grandparents bowling on Monday night. They are both over 65 and have their own bowling shoes. Lucas and his grandparents each plan to bowl 2 games. If Lucas pays for his shoes and for all of the games, how much will it cost?

2. Josh and his parents went bowling Monday night. They each bowled 4 games. His parents paid for the games and 3 pairs of shoes. How much did it cost? (Hint: Remember that Josh is a 19-year-old student.)

3. Josh went bowling on Saturday night. He paid the student price for 4 games and rented shoes. How much would he save by bowling on Monday night?

4. Movie tickets cost $7.50 on Saturday night and $5.25 on Tuesday night. How much do 4 friends save by going to a 7 P.M. movie on Tuesday instead of on Saturday?

5. **CRITICAL THINKING** Emily took 2 friends to a movie on Saturday night. The tickets cost $7.50 each. On Saturday afternoon the tickets cost only $3.50 each. How many friends could Emily have taken to the movie in the afternoon without spending more than the total cost of the tickets on Saturday night?

Maintaining Skills

Estimate each sum or product.

1. $47
   + 32
	2. $27
	   × 8
		3. $18.95
		   + 23.72

4. $27.00
   12.75
   + 11.50
	5. $5.75
	   × 9
		6. $12.95
		   35.89
		   + 6.75

17·3 How Can Comparing Prices Save You Money?

You decide to join a gym to get more exercise. There are three gyms in your town. They are the Golden Days Gym, the Fitness Fun Gym, and the Bright Lights Gym.

Read the chart below and compare the prices.

Gym Membership Costs	
Golden Days Gym	$336 per year
Fitness Fun Gym	$70 per month
Bright Lights Gym	$15 per day

▶ **EXAMPLE**

You plan to go to the gym 3 times a week for one year. Which gym would cost the least amount of money?

STEP 1 Find the yearly cost for Golden Days Gym.

$336 per year

STEP 2 Find the yearly cost for Fitness Fun Gym. Multiply the monthly cost by 12.

$70
× 12
$840 per year

Consumer Beware!
Many gyms offer trial memberships and discount rates. Sometimes if you join with a friend or relative, you can get a better rate. Be careful. Sometimes the rate is for a limited time.

STEP 3 Find the yearly cost for the Bright Lights Gym. First, multiply the daily cost by 3 to find the weekly cost.

$15
× 3
$45 per week

STEP 4 Then, multiply the weekly cost by 52 to find the yearly cost for Bright Lights Gym.

$45
× 52
$2,340 per year

STEP 5 Compare costs.

$336 < $840 < $2,340

Golden Days Gym would cost the least amount of money.